Lecture Notes in Mathematics

Edited by A. Dold and B. Eckmann

1090

Differential Geometry of Submanifolds

Proceedings of the Conference
held at Kyoto, January 23–25, 1984

Edited by K. Kenmotsu

Springer-Verlag
Berlin Heidelberg New York Tokyo 1984

Editor

K. Kenmotsu
Department of Mathematics
College of General Education, Tôhoku University
Kawauchi, Sendai 980, Japan

AMS Subject Classification (1980): 32E05, 32E10, 32F30, 35J20, 35J50, 49F10, 53A07, 53A10, 53B25, 53C35, 53C40, 53C42, 53C45

ISBN 3-540-13873-0 Springer-Verlag Berlin Heidelberg New York Tokyo
ISBN 0-387-13873-0 Springer-Verlag New York Heidelberg Berlin Tokyo

Library of Congress Cataloging in Publication Data. Main entry under title: Differential geometry of submanifolds. (Lecture notes in mathematics; 1090) "Conference on Differential Geometry of Submanifolds at the Research Institute for Mathematical Sciences, Kyoto University, January 23–25, 1984"–Pref. Includes index. 1. Geometry, Differential–Congresses. 2. Submanifolds–Congresses. I. Kenmotsu, K. (Katsuei), 1942-. II. Conference on Differential Geometry of Submanifolds (1984: Research Institute for Mathematical Sciences, Kyoto University) III. Series: Lecture notes in mathematics (Springer-Verlag); 1090. QA3.L28 no. 1090 510 s [516.3'6] 84-22129 [QA641]
ISBN 0-387-13873-0 (U.S.)

Printing and binding: Beltz Offsetdruck, Hemsbach/Bergstr.
2146/3140-543210

PREFACE

These proceedings report all the substance of talks contributed by participants in the Conference on Differential Geometry of Submanifolds at the Research Institute For Mathematicl Sciences, Kyoto University January 23-25, 1984.

The aims of the Conference was to present recent researches of young Japanese mathematicians which treated topics mentioned in the title and to stimulate discussions for future studies. In fact, contributors: T.Adachi, Y.Ohnita, S.Takakuwa and M.Koiso are students of graduate schools. I selected speakers of this conference under the suggestion of young members of our seminar in Tohoku University. Almost all included papers show very recent results of participants which were probably achieved in 1983.

We wish to thank all contributors for their eager participation and their collaboration in sending their texts. We would also like to acknowledge the financial support of the Research Institute for Mathematical Sciences, Kyoto University and Springer-Verlag for publishing this volume.

We have included as an appendix the text of a talk given subsequently at a conference in Tsukuba (Japan) in February, 1984, as it seemed a particulary apt addition to these proceedings.

Sendai, July 1984

Katsuei Kenmotsu

ESTIMATES FOR SOLUTIONS OF POISSON EQUATIONS AND THEIR APPLICATIONS TO SUBMANIFOLDS

Atsushi Kasue

Department of Mathematics
Osaka University, Toyonaka
Osaka, 560/ Japan

In this note, we shall first study some relations between the inner and the outer distances on a certain submanifold, making use of geometric estimates for solutions of Poisson equations. In Section 2, the Green potential of scalar curvature will be considered and a gap theorem for minimal submanifolds of Euclidean space will be given.

1. Some relations between the inner and the outer distances on a submanifold

Let M be a connected, complete Riemannian manifold of dimension m. We write Δ_M for the Laplace operator acting on functions (i.e., Δ_M = div. grad.). Given a nonnegative smooth function f on M and a domain $D \subset M$ with boundary ∂D, we consider the Poisson equation :

$$(1.1)\qquad \begin{aligned} \Delta_M u + f &= 0 \quad \text{on } D, \\ u &= 0 \quad \text{on } \partial D. \end{aligned}$$

Then the least nonnegative solution u of equation (1.1) (if it exists) can be represented by

$$u(x) = \int_D G_D(x,y)\, f(y)\, dy,$$

where $G_D(x,y)$ stands for the Green function of D, subject to the boundary condition $G_D = 0$ on ∂D.

We shall now consider the case D is a metric ball $B_M(x_o;r)$ around a point x_o of M with radius r. Let us take two continuous functions $R(t)$ and $f_*(t)$ on $[0,\infty)$ satisfying

(1.2) the Ricci curvature of M at a point $x \in M \geqq (m-1)\, R(\mathrm{dis}_M(x,x_o))$,

$f(x) \geqq f_*(\mathrm{dis}_M(x,x_o))$,

where $\mathrm{dis}_M(x,y)$ denotes the distance between two points x and y of M. Then the solution u of equation (1.1) on $B_M(x_o:r)$ has the following lower estimate :

$$(1.3)\qquad u(x) \geqq \int_{\mathrm{dis}_M(x,x_o)}^{r} \frac{\int_0^t F_R^{m-1}(s)\, f_*(s)\, ds}{F_R^{m-1}(t)}\, dt$$

(cf. [8: lemma (3.10)]), where $F_R(t) \in C^2([0,\infty))$ is the solution of equation :

$$(1.4)\qquad F_R''(t) + R(t)\, F_R(t) = 0 \ , \text{ with } F_R(0) = 0 \text{ and } F_R'(0) = 1 \ .$$

In order to obtain an upper bound for the solution u as above, we assume that M is a complete submanifold isometrically immersed into a complete Riemannian manifold $\overline{M}$. Initially, we treat the case : $f \equiv 1$. Given a point $\overline{x}_o$ of $\overline{M}$ and a positive number $\overline{r}$, suppose that the sectional curvature of $\overline{M}$ on $B_{\overline{M}}(\overline{x}_o:\overline{r}) \leqq$ a constant K, the injectivity radius of $\overline{M}$ at $\overline{x}_o > \overline{r}$, and moreover $\overline{r} \leqq \pi/2\sqrt{K}$ if $K > 0$. (We call such a metric ball <u>normal</u>.) Then it turns out from the well known comparison theorem that

$$\Delta_M \overline{\rho}^2 \geqq \begin{cases} 2\overline{\rho}\ [\ m(\log F_K)'\circ\overline{\rho} - |\, H_M \,|\ ] & (\ K > 0\ ) \\ 2\overline{\rho}\ [\ \dfrac{m}{\overline{\rho}} - |\, H_M \,|\ ] & (\ K = 0\ ) \\ 2\overline{\rho}\ [\ (m-1)(\log F_K)'\circ\overline{\rho} + \dfrac{1}{\overline{\rho}} - |\, H_M \,|\ ] & (\ K < 0\ ) \end{cases}$$

on $M \cap B_{\overline{M}}(\overline{x}_o:\overline{r})$, where $\overline{\rho}(x) = \mathrm{dis}_{\overline{M}}(x,\overline{x}_o)$, F_K is the solution of (1.4) with $R(t) = K$, and H_M denotes the mean curvature normal of the submanifold M . Let us assume here that

$$(1.5)\qquad \sup_{M \cap B_{\overline{M}}(\overline{x}_o:\overline{r})} |\, H_M \,| < C_o(m,K,\overline{r}) \ ,$$

where

$$C_o(m,K,\bar{r}) := \begin{cases} m\,(\log F_K)'(\bar{r}) & (K>0) \\ \dfrac{m}{\bar{r}} & (K=0) \\ (m-1)(\log F_K)'(\bar{r}) + \dfrac{1}{\bar{r}} & (K<0) \end{cases}$$

Then we have

$$\Delta_M \bar{\rho}^2 > 2\bar{r}\,[\,C_o(m,K,\bar{r}) - \sup_{M \cap B_{\bar{M}}(\bar{x}_o:\bar{r})} |\,H_M\,|\,] > 0$$

on $M \cap B_{\bar{M}}(\bar{x}_o:\bar{r})$. Therefore by the maximum principle, the least positive solution u of equation (1.1) ($f = 1$) has the following upper estimate :

$$\text{(1.6)} \qquad u(x) \leqq \frac{\bar{r}^2 - \bar{\rho}^2(x)}{2\bar{r}\,[\,C_o(m,K,\bar{r}) - \sup_{M \cap B_{\bar{M}}(\bar{x}_o:\bar{r})} |\,H_M\,|\,]}$$

for any $x \in M \cap B_{\bar{M}}(\bar{x}_o:\bar{r})$. Thus we obtain by (1.3) and (1.6)

$$\text{(1.7)} \qquad \int_0^r \frac{\int_0^t F_R^{m-1}(s)\,ds}{F_R^{m-1}(t)}\,dt < \frac{\bar{r}^2}{2\bar{r}\,[\,C_o(m,K,\bar{r}) - \sup_{M \cap B_{\bar{M}}(\bar{x}_o:\bar{r})} |\,H_M\,|\,]}$$

if $B_M(x_o:r) \subset M \cap B_{\bar{M}}(\bar{x}_o:\bar{r})$.

Lemma 1. Let $R(t)$ be a continuous function satisfying (1.2) and F_R the solution of equation (1.4). Set $\Phi(r)$ for the left side of (1.7). Then :

(i) $\Phi(r) \geqq C_1\,r^2$ if $R(r) \geqq -\dfrac{C_2}{r^2}$.

(ii) $\Phi(r) \geqq C_3\,r$ if $R(r) \geqq -C_4$.

(iii) $\Phi(r) \geqq C_5 \log^{J+1} r$ if $R(t) \geqq -C_6[\,r \prod_{j=0}^{J} \log^j r\,]^2$.

Here C_i $(i = 1,\ldots,6)$ are positive constants, $\log^2 r := \log(\log r),\ldots,$ $\log^j r := \log^{j-1}(\log r)$, and the last two inequalities are assumed to hold for sufficiently large r .

Proof. The above three assertions will be derived from simple computation. Let us prove here the last one. Suppose $R(r) \geqq -C_6[r \prod_{j=0}^{J} \log^j r]^2$ for any $r \geqq C_7 > 0$. Choose a positive constant C_8 such that $C_8^2 > C_6$ and $C_8[C_7 \prod_{j=0}^{J} \log^j C_7] > (\log F_R)'(C_7)$. Set $a(r) := C_8[r \prod_{j=0}^{J} \log^j r]$ ($r \geqq C_7$) and $A(r) := \exp \int_{C_7}^{r} a(s)\, ds$. Then we have

$$(A F_R' - A' F_R) \Big|_{t=C_7}^{t=r} = \int_{C_7}^{r} A(t)F_R''(t) - A''(t)F_R(t)\, dt$$

$$= \int_{C_7}^{r} (-R(t) - a^2(t) - a'(t))A(t)F_R(t)\, dt$$

$$< 0 .$$

This implies that $(\log F_R)'(r) \leqq a(r)$ for any $r \geqq C_7$, so that

$$\frac{\int_0^r F_R^{m-1}(s)\, ds}{F_R^{m-1}(r)} \geqq \frac{C_9}{a(r)}$$

for every $r > C_{10}$, where C_9 and C_{10} are some positive constants. Thus we obtain

$$\Phi(r) \geqq C_9 \int_{C_7}^{r} \frac{dt}{a(t)} = C_9[\prod_{j=0}^{J} \log^j r - \prod_{j=0}^{J} \log^j C_7] .$$

This proves the last assertion of the lemma.

The following two propositions are immediate consequences of inequality (1.7) and the preceding lemma.

Proposition 1. Let M be a complete Riemannian manifold of dimension m isometrically immersed into a complete Riemannian manifold $\bar{M}$. Suppose that the scalar curvature of M is bounded from below by $-C[\, dis_M(*,x_o) \prod_{j=0}^{J} \log^j dis_M(*,x_o)\,]^2$ outside a compact set, where C

is a positive constant, J is a positive integer and x_o is a fixed point of M. Then if M is contained in a closed normal metric ball $B_{\overline{M}}(\bar{r})$ of $\overline{M}$ with radius $\bar{r}$ one has

$$\sup_M |H_M| < \begin{cases} m\sqrt{K}\,\mathrm{cotan}\sqrt{K}\,\bar{r} & (K > 0) \\ \dfrac{m}{\bar{r}} & (K = 0) \\ (m-1)\sqrt{-K}\,\mathrm{cotanh}\sqrt{-K}\,\bar{r} + \dfrac{1}{\bar{r}} & (K < 0). \end{cases}$$

Here H_M (resp. K) denotes the mean curvature normal of the submanifold M (resp. the supremum of the sectional curvature of $\overline{M}$ on $B_{\overline{M}}(\bar{r})$).

<u>Proposition</u> 2. Let M be a complete minimal submanifold immersed into a complete, simply connected Riemannian manifold $\overline{M}$ of nonpositive curvature. Fix a point x_o of M and set $D(r) := \max\{ \mathrm{dis}_{\overline{M}}(x_o, x) : x \in \partial B_M(x_o, r)\}$. Then:

(i) $D(r) \geqq C_1 r$ if the scalar curvature of $M \geqq -C_2 \mathrm{dis}_M(x_o,*)^{-2}$.

(ii) $D(r) \geqq C_3 r^{1/2}$ if the scalar curvature of $M \geqq -C_4$.

(iii) $D(r) \geqq C_5 \prod_{j=0}^{J+1} \log^j r$ (for large r) if the scalar curvature of $M \geqq -C_6 [\mathrm{dis}_M(x_o,*) \prod_{j=0}^{J} \log^j \mathrm{dis}_M(x_o,*)]^2$ outside a compact set.

In particular, the image of M is unbounded in $\overline{M}$.

Here C_i $(i=1,\dots,6)$ are positive constants and J is a positive integer.

By the Gauss equation, we have the following

<u>Corollary to Proposition</u> 1. Let M, $B_{\overline{M}}(\bar{r})$, K and $\bar{r}$ be as in Proposition 1. Set k for the minimum of the sectional curvature of $\overline{M}$ on $B_{\overline{M}}(\bar{r})$. Then if the codimension of M is equal to one, one has

$$\sup_{\substack{X \in T_xM \\ x \in M}} \mathrm{Ric}_M(X) \geq \begin{cases} (m-1)[\sqrt{K}(\mathrm{cotan}\,\sqrt{K}\,\bar{r})^2 + k\,] & (K > 0) \\ (m-1)[\,\dfrac{1}{\bar{r}^2} + k\,] & (K = 0) \\ (m-1)[\,\dfrac{1}{m^2}\,((m-1)\,\sqrt{-K}\,\mathrm{cotanh}\,\sqrt{-K}\,\bar{r} + \dfrac{1}{\bar{r}})^2 + k] & \\ & (K < 0), \end{cases}$$

where $\mathrm{Ric}_M(X)$ stands for the Ricci curvature of M in the direction of $X \in T_xM$.

We shall now consider the special case M is a complete Riemannian submanifold immersed into Euclidean space R^n. For any nonnegative integer $p : 0 \leq p \leq n-2$ and a positive number $\bar{r}$, set $C(p:\bar{r}) := \{ (x_1,\ldots,x_n) \in R^n : |x_1|^2 + \cdots + |x_{n-p}|^2 \leq \bar{r}^2 \}$. Suppose that

$$m \geq p+1 ,$$

$$\sup_{M \cap C(p:\bar{r})} |\, H_M \,| < \frac{1}{\bar{r}} .$$

Then by the same calculation which has led inequality (1.7), we obtain

$$(1.8) \qquad \int_0^r \frac{\int_0^t F_R^{m-1}(s)\,ds}{F_R^{m-1}(t)}\,dt < \frac{\bar{r}^2}{2[\,1 - \bar{r} \sup_{M \cap C(p:r)} |\, H_M \,|]}$$

if $B_M(x_o:r) \subset C(p:\bar{r})$. This inequality and Lemma 1 (iii) show the following

Proposition 3. Let M be a complete Riemannian manifold of dimension m isometrically immersed into Euclidean space R^n. Suppose the scalar curvature of M is bounded below by $-C[\mathrm{dis}_M(x_o,*) \prod_{j=0}^{J} \log^j \mathrm{dis}_M(x_o,*)]^2$ outside a compact set, where C is a positive constant and J is a positive integer. Then if M is contained in a "cylindrical" domain $C(p:\bar{r})$ of R^n and $m \geq p+1$, one has

$$\sup_M |\, H_M \,| \geq \frac{1}{\bar{r}} .$$

Here H_M denotes the mean curvature normal of the submanifold M.

Corollary. Let M and $\bar{r}$ be as in Proposition 3. Then if $m = n-1$,

$$\sup_{\substack{X \in T_x M \\ x \in M}} \mathrm{Ric}_M(X) \geqq \frac{m-1}{m^2 \bar{r}^2} .$$

Moreover by virtue of a result due to Schoen [11: Corollary 4], we have another application of inequality (1.7) as follows:

Theorem 1. Let Σ be a complete stable minimal surface with boundary $\partial\Sigma$ immersed into Euclidean space R^3. Suppose the boundary $\partial\Sigma$ is contained in a cylinder of R^3 with radius $\bar{r}$. Then one has

$$\sup_{x \in \Sigma} \mathrm{dis}_\Sigma(x, \partial\Sigma) < C_1 \bar{r}$$

$$\sup_{x \in \Sigma} \mathrm{Vol.}(B_\Sigma(x, \mathrm{dis}\ (x, \partial\Sigma)/2)) < C_2 \bar{r}^2 ,$$

where C_1 and C_2 are absolute positive constants.

Remarks. (1) Proposition 1 is a generalization of a theorem by Jorge and Xavier [6] . Their method does not seem to yield our results (espe cially, Proposition 2 and Theorem 1), since a generalized maximum principle due to Omori [10] plays an essential role there. The reader will be also able to find other related results and some references in, e.g., [5]. (2) The hypothesis on the scalar curvature of the submanifold can not be omitted from the results of this section (except Theorem 1). In fact, it is known that there exist complete, bounded, minimal submanifolds of Euclidean space (cf. [4]). (3) Let $\phi: M \to \bar{M}$ be a harmonic map from a complete Riemannian manifold M into a complete Riemannian manifold $\bar{M}$. We write $e(x)$ for the energy density of ϕ. Suppose the closure of the image $\overline{\phi(M)}$ is compact and there exists a strictly convex function η defined on a neighborhood of $\overline{\phi(M)}$. Then $\Delta_M \eta \circ \phi(x) = \mathrm{trace}\ \phi^* (\bar{\nabla}^2 \eta)(x) \geqq c\, e(x)$ ($c > 0$) on M and hence, by the maximum principle, we have

$$\int_M G_M(x,y)\, e(y)\, dy < +\infty$$

on M. On the other hand, it follows from (1.3) ($r = +\infty$) that

$$(1.9) \qquad \int_0^\infty \frac{\int_0^t F_R^{m-1}(s)\, e_*(s)\, ds}{F_R^{m-1}(t)}\, dt \leqq \int_M G_M(x_o,y)\, e(y)\, dy$$

where x_o is a fixed point of M, F_R is as in (1.3) and $e_*(t) := \min \{ e(x) : x \in M,\ dis_M(x_o,x) = t \}$. Therefore we see that if $\overline{M}$ possesses a strictly convex function and the right side of (1.9) is infinite (e.g., $e_*(t) \geqq C_1 > 0$, $R(t) \geqq -C_2[\, t \prod_{j=0}^{J} \log^j t\,]^2$ for large t, or $e_*(t) \geqq C_3/t$, $R(t) \geqq -C_4$ for large t), $\phi(M)$ must be unbounded in $\overline{M}$. This is a generalization of the last assertion in Proposition 2.

2. Green potential of scalar curvature

In this section, we shall show an upper bound for the solution of equation (1.1) under certain conditions and state a gap theorem for minimal submanifolds of Euclidean space.

Let M be a complete minimal submanifold of dimension m immersed into a complete, simply connected Riemannian manifold $\overline{M}$ of nonpositive curvature. Given a nonnegative smooth function f on M and a point x_o of M, we assume that there exists a continuous function $f^*(t)$ on $[0,\infty)$ satisfying

$$(2.1) \qquad 0 \leqq f(x) \leqq f^*(\, dis_M(x,x_o)\,) \qquad \text{on } M,$$

$$(2.2) \qquad \frac{m-1}{t^m} \int_0^t s^{m-1} f^*(s)\, ds \geqq C f^*(t) \qquad \text{on } [0,\infty)$$

for some positive constant $C < 1$. In what follows, we write $\overline{\rho}(x)$ ($x \in M$) and $M_{\overline{r}}$, respectively, for <u>the distance on</u> $\overline{M}$ between two point x and x_o of M and the intersection of M and $B_{\overline{M}}(x_o : \overline{r})$. Set

$$\Psi_{\overline{r}}(t) := \int_t^{\overline{r}} \frac{\int_0^s v^{m-1} f^*(v)\, dv}{s^{m-1}}\, ds \qquad (\overline{r} > 0).$$

Then we have

$$\Delta_M \Psi_{\overline{r}} \circ \rho \leqq - C f^* \circ \rho \qquad \text{on } M$$

and

$$\Psi_{\bar r} \circ \bar\rho = 0 \qquad \text{on} \quad \partial M_{\bar r} \ ,$$

since $\Delta_M \bar\rho \geqq (m-1)/\bar\rho$. The maximum principle implies that the least positive solution $u_{\bar r}(x)$ of equation (1.1) on $M_{\bar r}$ has an estimate:

(2.3) $$C\, u_{\bar r}(x) \leqq \Psi_{\bar r} \circ \bar\rho$$

on $M_{\bar r}$. In particular, we see that

(2.4) $$C\, u_{\bar r}(x_o) \leqq \Psi_{\bar r}(0) \ .$$

As immediate consequences of (2.3) and (2.4), we obtain the following assertions :

(2.5) $\int_M G_M(x,y)\, f(y)\, dy \leqq C_1/(1+\rho(x))^{\varepsilon}$ on M $(\varepsilon > 0)$ if $m \geqq 3$ and $f(x) \leqq C_2/(1+\bar\rho(x))^{2+\varepsilon}$ on M .

(2.6) $\int_{M_{\bar r}} G_{\bar M}(x_o,y)\, f(y)\, dy \leqq C_3 \log(1+\bar r)$ for $\bar r > 0$ if $f(x) \leqq C_4/(1+\bar\rho(x))^2$ on M .

(2.7) $\int_{M_{\bar r}} G_{\bar M}(x_o,y)\, f(y)\, dy \leqq C_5\,(1+\bar r)^{k+2}$ for $\bar r > 0$ if $k > -2$ and $f(x) \leqq C_6\,(1+\bar\rho(x))^k$ on M .

We are now interested in the case $M = R^n$ and $f(x) = |$ the scalar curvature of M at $x \in M\, |$. Actually, we can derive a gap theorem below from the first assertion (2.5) :

<u>Theorem</u> 2. Let M be a complete minimal submanifold of dimension m immersed into Euclidean space R^n . Suppose that

(i) $m \geqq 3$,

(ii) the scalar curvature of M at $x \in M \geqq -\dfrac{C}{1+|x|^{2+\varepsilon}}$

($C > 0$, $\varepsilon > 0$) .

Then M must be an m-plane of R^n, provided that

(#) M possesses no bounded harmonic functions except constants.

The proof will be given in the forthcoming paper [9].

Remarks. (1) Condition (i) can not be omitted from Theorem 2 (cf. [9: Example 1]). (2) The positive constant ε in condition (ii) is also necessary for Theorem 2. In fact, Vitter [12] showed that a 'generic' algebraic hypersurface of C^n has the sectional curvature uniformly bounded in absolute value by $C/|z|^2$ ($C > 0$). (3) In case M is a complex submanifold properly imbedded into C^n, Theorem 2 turns out to be true without condition (#) (cf. [9: Theorem A]). (4) In case M is an area-minimizing hypersurface properly imbedded into R^n, Bombieri and Giusti [1] proved that M satisfies condition (#), so that Theorem 2 holds for such a hypersurface without condition (#). (5) When the constant C in assumption (ii) is less than $C(\varepsilon) := (1+2/\varepsilon)(\varepsilon/2)^{-2/(2+\varepsilon)} = \min\{(1+t^{2+\varepsilon})t^{-2} : t > 0\}$, condition (#) can be derived from assumption (ii) (cf. [9]). (6) Let M be a complete minimal submanifold immersed into R^n. Suppose that $m \geqq 3$ and the scalar curvature $S_M(x)$ of M at $x \in M \geqq - C/|x|^2$. Then M is stable, provided that $C \leqq (m-2)^2/4$. In fact, a positive function $G(x)$ on M defined by $G(x) := (m-1)^{-1/2}|x|^{(2-m)/2}$ satisfies

$$\Delta_M G(x) + |S_M(x)|G(x) \leqq 0 ,$$

so that M is stable (cf. [3]). (7) Suppose that M is a complex submanifold of C^n defined by the system of equations $F = (F_1,\dots,F_p) = (0,\dots,0)$ ($p = n - m$) and suppose that the defferential dF is surjective everywhere on M. Then the Ricci form Ric_M of M is given by $Ric_M = - [\sqrt{-1}\,\partial\bar\partial \log J[F]^2]_{|M}$ (the restriction of $\sqrt{-1}\,\partial\bar\partial \log J[F]^2$ to M), where

$$J[F] := \left(\sum_{1 \leqq i_1 < \cdots < i_p \leqq n} | \det[\partial F_j/\partial z_{i_k}] |^2 \right)^{1/2}.$$

Consequently, the scalar curvature S_M of M is represented by

$$S_M = - \Delta_M \log J[F]^2{}_{|M} .$$

Observe that

$$(2.8)\qquad \log J[F]^2(z) = \int_{M_{\bar r}} G_{M_{\bar r}}(z,w)\, S_M(w)\, dw + \int_{\partial M_{\bar r}} \log J[F]^2(\xi)\, P_{\bar r}(z,\xi)\, d\xi$$

on $M_{\bar r} := M \cap B_{C^n}(o:\bar r)$, where $P_{\bar r}(z,\xi)$ denotes the Poisson kernel $M_{\bar r}$. Let $\eta(t) > 0$ be an increasing function on $[0,\infty)$ such that $\log |F(z)| \leqq \eta(|z|)$. Then it follows from (2.8) that for some positive constant C

$$\int_{M_{\bar r}} G_{M_{\bar r}}(z_o,w)\, |S_M(w)|\, dw \leqq C\, \eta(\bar r)$$

for large $\bar r > 0$, where z_o is a fixed point of M. In particular, we see that

$$(2.9)\qquad \int_{M_{\bar r}} G_{M_{\bar r}}(z_o,w)\, |S_M(w)|\, dw \leqq C_1 \log(1 + \bar r) \qquad (C_1 > 0)$$

if each component of F is polynomial, and

$$(2.10)\qquad \int_{M_{\bar r}} G_{M_{\bar r}}(z_o,w)\, |S_M(w)|\, dw \leqq C_2\, (1 + \bar r)^{\tau+\varepsilon}$$

if each component F_i of F is a holomorphic function of finite order $\tau \geqq 0$. i.e., for any $\varepsilon > 0$. there exists a constant $C(\varepsilon) > 0$ such that $\log |F_i| \leqq C(\varepsilon)\, |z|^{\tau+\varepsilon}$. Note that we have estimate (2.9) (resp. (2.10)) when $S_M(z) \geqq - C/|z|^2$ (resp. $S_M(z) \geqq - C\,(1 + |z|)^{\tau-2+\varepsilon}$) (cf. (2.6), (2.7)).

3. Appendix

Before concluding this note, we shall prove a proposition concerning the Cauchy probrem for the heat equation on a certain Riemannian manifold, making use of the method in [2]. Our result is

<u>Proposition</u> 4. Let M be a complete Riemannian submanifold <u>properly</u> immersed into a complete Riemannian manifold $\bar M$.

(1) Suppose that the sectional curvature of $\bar M \geqq -C_1 [\bar\rho \prod_{j=0}^{J} \log^j \bar\rho\,]^2$

outside a compact set of $\bar{M}$ and the length of the mean curvature normal H_M of the submanifold $M \leqq C_2[\,\bar{\rho}|_M \prod_{j=0}^{\tilde{J}} \log^j \bar{\rho}|_M\,]^2$ outside a compact set of M, where C_1 and C_2 are positive constants, J and $\tilde{J}$ are positive integers, and $\bar{\rho}$ stands for the distance function on $\bar{M}$ to a fixed point $\bar{x}_o$ of $\bar{M}$. Then M has a unique fundamental solution $P_M(x,y,t)$ of the heat equation which satisfies

$$\int_M P_M(x,y,t)\,dy = 1$$

for any $x \in M$ and every $t > 0$.

(2) Suppose that the sectional curvature of M is bounded from below and the length of the mean curvature normal of M is bounded from above. Let u_o be a continuous function on M which vanishes at infinity. Then

$$P_t[u_o](x) := \int_M P_M(x,y,t)\,u_o(y)\,dy$$

vanishes at infinity for every $t > 0$.

<u>Proof</u>. It is enough to prove Proposition 4 in case M is noncompact, so that in what follows M is assumed to be noncompact. Let us take a continuous function $k(t) \geqq 0$ (resp. a continuous function $h(t) \geqq 0$) on $[0,\infty)$ such that the sectional curvature of $M \geqq k\circ\bar{\rho}$ (resp. $|H_M| \leqq h\circ\bar{\rho}|_M$). Let $F_k \in C^2([0,\infty))$ be the solution of equation (1.4) ($R = k$). Set

$$H(t) := \exp \int_0^t h(s)\,ds$$

$$\Phi(t) := \int_0^t \frac{\int_0^s F_k{}^m(v)\,H(v)\,dv}{F_k{}^m(s)\,H(s)}\,ds .$$

Suppose that for some positive constant $C : 0 < C < 1$,

$$[\,m\,(\log F_k)'(t) + h(t)\,] \int_0^t F_k{}^m(s)\,H(s)\,ds \geqq C\,F_k{}^m(t)\,H(t)$$

($t > 0$). Then we have

$$\Delta_M \Phi\circ\bar{\rho}|_M \leqq C$$

as a distribution on M (cf. [7 : Theorem (2.31)]). Thus the proposition is a consequece of Lemma 2 below and simple computation (cf. Lemma 1).

Lemma 2. Let M be a complete Riemannian submanifold properly immersed into a complete Riemannian manifold $\overline{M}$.

(1) If there exist an increasing function $\Phi(t) \in C^0([0,\infty))$ and a point $\overline{x}_0$ of $\overline{M}$ such that $\lim_{t \to \infty} \Phi(t) = +\infty$ and $\Delta_M \Phi \circ \overline{\rho}_{\overline{x}_0} |M \leq 1$ ($\overline{\rho}_{\overline{x}_0} := \mathrm{dis}_{\overline{M}}(\overline{x}_0, *)$) as a distribution on M, then the same conclusion as in Proposition 4 (1) holds.

(2) If there exists an increasing function $\Phi(t) \in C^0([0,\infty))$ such that $\lim_{t \to \infty} \Phi(t) = +\infty$ and for any point $\overline{x}_0 \in \overline{M}$, $\Delta_M \Phi \circ \overline{\rho}_{\overline{x}_0} |M \leq 1$ as a distribution on M, then the same conclusion as in Proposition 4 (2) holds to be true.

Proof. This lemma can be derived from the argument in [2].

Note that the properness of the immersion is necessary for Proposition 4. Actually, let M be, for example, a complete, bounded minimal submanifold immersed into R^n (cf. Remark (2) in Section 1). Then there exists, on M, a positive solution of equation (1.4) with $f = 1$ (cf. (1.6)). That is, $\int_M G_M(x,y)\, dy < +\infty$ for any $x \in M$. This implies that $\int_M P_M(x,y,t)\, dy < 1$ for some $t > 0$, because $G_M(x,y) = \int_0^\infty P_M(x,y,t)\, dt$, where $P_M(x,y,t)$ is the minimal positive fundamental solution of the heat equation on M. Finally, we remark that the lower bound for the sectional curvature of M in Proposition 4 (1) is optimum for the conclusion (cf. [8 : Section 5]).

References

[1] E. Bombieri and E. Giusti : Harnack's inequality for elliptic differential equations on minimal surfaces, Inven. math. 15 (1972) 24 - 46.

[2] J. Dodzuik : Maximum principles for parabolic inequalities and the heat flow on open manifolds, Indiana Univ. Math. J. 32 (1983), 703 -716.

[3] D. Fischer-Corbrie and R. Schoen : The structure of complete stable minimal surfaces, Comm. Pure Appl. Math.33 (1980), 199 - 211.

[4] P.W. Jones : A complete bounded complex submanifold of C^3, Proc. A.M.S. 76 (1979), 305 - 306.

[5] L. Jorge and D. Koutroufiotis : An estimate for the curvature of bounded submanifolds, Amer. J. Math. 103 (1981), 711 - 725.

[6] L. Jorge and F.V. Xavier : An inequality between the exterior diameter and the mean curvature of bounded immersions, Math. Z. 178 (1981), 77 - 82.

[7] A. Kasue : A Laplacian comparison theorem and function theoretic properties of a complete Riemannian manifold, Japanese J. Math. (1982), 309 - 341.

[8] A. Kasue : Applications of Laplacian and Hessian comparison theorems, Advanced Studies in Pure Math. 3, Kinokuniya, Tokyo.

[9] A. Kasue : Gap theorems for minimal submanifolds of Euclidean space, to appear.

[10] H. Omori : Isometric immersions of Riemannian manifolds, J. Math. Soc. Japan 19 (1967), 205 - 214.

[11] R. Schoen : Estimates for stable minimal surfaces in three dimensional manifolds, Seminar on Minimal Submanifolds, Princeton U. Press (1983), 111 - 126.

[12] A. Vitter : On the curvature of complex hypersurfaces, Indiana Univ. Math. J. 23 (1974), 813 - 826.

TAUT EMBEDDINGS AND DUPIN HYPERSURFACES

REIKO MIYAOKA

Department of Mathematics
Tokyo Institute of Technology
Meguro, Tokyo 152 JAPAN

§0. Introduction.

The first step to investigate taut embeddings $M^n \subset S^{n+1}$ is to find as many examples as possible. The family of isoparametric hypersurfaces and their conformal images are known as typical examples [CR2]. They satisfy:

D1. The multiplicities of the principal curvatures are constant.

D2. Each principal curvature is constant along its curvature line.

An immersed hypersurface in a complete simply connected space form $\bar{M}^{n+1}$ satisfying D1 and D2 is called a Dupin hypersurface.

In 1983, Thorbergsson [T] proved that a compact embedded Dupin hypersurface in $\bar{M}^{n+1}$ is taut with respect to $\mathbb{Z}_2$ coefficient and that dim $H^*(M,\mathbb{Z}_2) = 2h$, where h is the number of principal curvatures which takes value $\in \{1,2,3,4,6\}$. This suggests a close relation between compact embedded Dupin hypersurfaces and isoparametric hypersurfaces. In fact, it is known [CR1] that compact embedded Dupin hypersurfaces are conformal images of isoparametric hypersurfaces for h = 1 and 2. But this is not true for $h \geqq 3$.

In 1981, Pinkall [P] showed that the class of Dupin hypersurfaces is invariant under Lie-transformations (cf.§2), bigger than conformal transformations. For instance, a parallel hypersurface of a (non-trivial) conformal image of an isoparametric hypersurface M with three principal curvatures is a Dupin hypersurface, which is a Lie-image of M but is not a conformal image of any isoparametric hypersurfaces. The author [M] proved that for h = 3, compact embedded Dupin hypersurfaces are Lie-images of isoparametric hypersurfaces.

On the other hand, Pinkall mentioned that taut embeddings $M^n \subset \bar{M}^{n+1}$ are also invariant under Lie-transformations. Note that there exist taut embeddings $M^n \subset \bar{M}^{n+1}$ which are not Dupin hypersurfaces. For instance, a tube M of an embedding $S^{k_1} \times \ldots \times S^{k_r} \subset E^{k_1+\ldots+k_r+r}$ is taut [CW], which satisfies D2 on M but dose not satisfy D1 on a measure-zero subset of M.

Respecting Pinkall's original definition of Dupin hypersurfaces, we call in this paper, a hypersurface satisfying D2 a P-Dupin hypersurface.

Now, we may ask: Are there any relation between taut embeddings and P-Dupin hypersurfaces?

Our main result is:

Theorem I. Let $M^n \subset \bar{M}^{n+1}$ be a taut embedding of a smooth compact connected manifold M. Then M is a P-Dupin hypersurface.

Note that on an open subset U of M where D1 is satisfied, D2 holds automatically for all principal curvatures with multiplicities $\geqq 2$, without any assumption on M. Tautness guarantees D2 for simple principal curvatures, too, and this is our result.

In the forth-coming paper, using this theorem, we classify taut embeddings $M^3 \subset \bar{M}^4$ as follows:

Theorem II. Let $M^3 \subset \bar{M}^4$ be a taut embedding of a smooth compact connected manifold M. Then M is a Lie-image of one of the following hypersurfaces:

(a) $S^3(r)$ in $S^4(1)$, $0<r<1$.

(b) $S^2(r)\times S^1(\sqrt{1-r^2})$ in $S^4(1)$, $0<r<1$.

(c) An isoparametric hypersurface in $S^4(1)$ with three principal curvatues (= a tube of the Veronese surface in $S^4(1)$).

(d) A tube in $S^4(1)$ of a cyclide of Dupin in $S^3(1)$.

Corollary. Let $M^3 \subset \bar{M}^4$ be as above. Then $\mathbb{Z}_2$-homology of M is one of the following: For $k = 0,3$ and $\ell = 1,2$,

(a) $H_k(M,\mathbb{Z}_2) = \mathbb{Z}_2$, $H_\ell(M,\mathbb{Z}_2) = 0$.

(b) $H_k(M,\mathbb{Z}_2) = H_\ell(M,\mathbb{Z}_2) = \mathbb{Z}_2$.

(c) $H_k(M,\mathbb{Z}_2) = \mathbb{Z}_2$, $H_\ell(M,\mathbb{Z}_2) = \mathbb{Z}_2 \oplus \mathbb{Z}_2$.

(d) $H_k(M,\mathbb{Z}_2) = \mathbb{Z}_2$, $H_\ell(M,\mathbb{Z}_2) = \mathbb{Z}_2 \oplus \mathbb{Z}_2 \oplus \mathbb{Z}_2$.

Besides the known result in case n = 2 (cf.§1), we answer Kuiper's question [CR2] affirmatively for n = 2 and 3; i.e. taut embeddings $M^n \subset \bar{M}^{n+1}$ are algebraic for n = 2 and 3.

This paper consists of five sections, containing a brief introduction of taut embeddings and Lie-transformations in §§1-2. We prove Theorem I in §3, and in §4, some ideas of the proof of Theorem II will be described. We finish the paper by giving open problems concerning taut embeddings.

§1. Taut embeddings and examples.

Let $f: M^n \to E^N$ be an immersion of a smooth compact connected manifold into Euclidean space. For any $z \in E^N$, we define

$H_z : M \to R$ by $H_z(p) = \langle f(p), z \rangle$,

$L_z : M \to R$ by $L_z(p) = |f(p)-z|^2$,

where $\langle\ ,\ \rangle$ is the Euclidean inner product.

The immersion f is called <u>tight</u>, respectively <u>taut</u>, if there exists a field $\mathbb{F}$ such that any non-degenerate function H_z, respectively L_z, has the minimum number of critical points required by the Morse inequalities with respect to $\mathbb{F}$. We say that f is <u>spherical</u> if the image of f lies in a Euclidean hypersphere $S^{N-1} \subset E^N$.

Following facts are known [CW]:

<u>Fact 1.1</u>

(a) If f is taut, then f is tight.

(b) If f is spherical tight, then f is taut.

(c) If f is taut, then f is an embedding.

(d) Tightness is projectively invariant.

(e) Tautness is conformally invariant.

As for an example, a tight embedding of a topological sphere is the boundary of a convex body [CL], while its taut embedding is a Euclidean sphere [NR].

Note that the codimension is restricted as follows [K]:

<u>Fact 1.2</u> If f is tight and full, then $N \leq \frac{1}{2}n(n+3)$.

From this, Fact 1.1 (a) and (e), we get

<u>Fact 1.3</u> If f is taut and full, then $N \leq \frac{1}{2}n(n+3)$. If the equality holds, then f is spherical.

In this paper, we treat taut embeddings $M^n \subset \bar{M}^N$, where $\bar{M}^N = E^N$, S^N or H^N. In each case, tautness is defined similarly by using the distance function $L_z = d(z,\ )^2$, where $d(z,\ )$ denotes the distance in $\bar{M}^N$ from $z \in \bar{M}^N$. Since stereographic projections are conformal, investigation of taut embedding M^n into E^N corresponds to that of M^n into S^N or into H^N automatically.

Now, we give some examples of taut embeddings $M^n \subset E^N$.

<u>Example 1</u>. For surface case $n = 2$, N should be 3,4 and 5. All taut surfaces are classified as follows, where π means any stereographic projection $S^N \to E^N$ from a point of $S^N \setminus M$.

E^N	E^3	E^4	E^5
M^2	$S^2(r)$	$S^1(r)\times S^1(\sqrt{1-r^2})$ $(\subset S^3)$ and its conformal images	Veronese surfaces $(\subset S^4)$ [K]
	π: Cyclides of Dupin [B]	π: Stereographic images of Veronese surfaces [CW]	

Example 2. For N = n+1, any isoparametric hypersurfaces are taut. Note that they are all algebraic, while some of them are non-homogeneous [OT].

Example 3. Focal sets of isoparametric hypersurfaces are taut [CR2] (codim>1).

Example 4. Takeuchi-Kobayashi [TK] proved that the standardly embedded symmetric R-spaces are spherical tight, so they are taut (Generally, the codimensions are rather big).

Example 5. Tubes of certain taut embeddings $M^n \subset E^N$ (cf. [CW]).

Note that Lie-images (cf. §2), especially conformal images of all examples are also taut.

§2. Lie-transformations.

Pinkall first pointed out that the class of P-Dupin hypersurfaces is invariant under Lie-transformations. Moreover he says that tautness is preserved by Lie-transformations. Therefore, to treat these subjects, we must consider them from the standpoint of Lie's sphere geometry.

As a definition, a Lie-transformation of $\overline{M}^N$ is a transformation on the set $\mathcal{S}$ = {all oriented hyperspheres of $\overline{M}^N$} preserving oriented contact of any two elements of $\mathcal{S}$. Note that $\mathcal{S}$ contains all points of $\overline{M}^N$ and all hyperplanes of $\overline{M}^N$ as special hyperspheres.

We give here a brief introduction of Lie-transformations of $\overline{M}^N = S^N$. For convenience, let S^N be the unit sphere in E^{N+1} with center at the origin. Extend E^{N+1} to the projective space P^{N+1} by identifying $x \in E^{N+1}$ with $(1,x) \in P^{N+1}$ using the homogeneous coordinates of P^{N+1}.

Now, a hypersphere $S \in \mathcal{S}$ lies on a hyperplane H_S of E^{N+1} which determines the pole $k \in P^{N+1}$ with homogeneous coordinates $(1,k_0)$. Using symmetric bilinear form σ with signature $(-,+,\dots,+)$ in V^{N+2} (the vector space associated with P^{N+1}), we know that $\sigma(k,k) = |k_0|^2-1 > 0$ if S is not a point of S^N. In order to distinguish the orientation of S, extending P^{N+1} to P^{N+2} by identifying $k \in P^{N+1}$ with $(k,1) \in P^{N+2}$ using the homogeneous coordinates, we obtain a mapping $\psi: \mathcal{S} \to P^{N+2}$ such that

$\psi: \mathcal{S} \ni S \longmapsto \tilde{k} = (k, \pm\sqrt{\sigma(k,k)}) \in P^{N+2}$.

Now, let $\tilde{\sigma}$ be the symmetric bilinear form in V^{N+3} with signature $(-,+,\ldots,+,-)$ and let $Q = \{\tilde{k} \in P^{N+2} | \tilde{\sigma}(\tilde{k},\tilde{k}) = 0\}$. Then it is not difficult to see that ψ gives a one-to-one correspondence between $\mathcal{S}$ and Q. Thus we identify $\mathcal{S}$ with Q. An oriented contact of two hyperspheres $\tilde{k}_1$ and $\tilde{k}_2$ in Q is represented by $\tilde{\sigma}(\tilde{k}_1,\tilde{k}_2) = 0$.

Definition 1. Möbius transformations of S^N is the projective transformations on P^{N+1} preserving σ. The group of Möbius transformations of S^N is $O(N+1,1)/\{I,-I\}$ and its dimension is $(N+2)(N+1)/2$. They consist of isometries and conformal transformations of S^N.

Definition 2. Lie-transformations of S^N is the projective transformations on P^{N+2} preserving $\tilde{\sigma}$. The group of Lie-transformations of S^N is $O(N+1,2)/\{I,-I\}$ and its dimension is $(N+3)(N+2)/2$. The Möbius transformation group is a subgroup of this group.

Now, what does a Lie-image of a hypersurface in S^N mean? A hypersurface M in S^N corresponds to a figure in Q as follows. At a point $p \in M$, there is a one-parameter family of oriented hyperspheres $S_p(t)$ tangent to M at p. Obviously $\tilde{\sigma}(\psi(S_p(t)),\psi(S_p(t'))) = 0$ for all $t,t' \in \mathbb{R}$. On the other hand, if a straight line in P^{N+2} expressed as

$$\tilde{k}(t) = \tilde{k}_1 + t\tilde{k}_2,\ \tilde{k}_1,\tilde{k}_2 \in V^{N+3}-\{0\},\ \text{linearly independent},\ t \in \mathbb{R},$$

lies in Q for all $t \in \mathbb{R}$, then we see that $\tilde{k}(t)$ determines a one-parameter family of oriented hyperspheres in S^N touching each other at a point of S^N.

In this way, a hypersurface M in S^N corresponds to the n-parameter family of straight lines of P^{N+2} contained in Q and vise versa. For details, see [P].

By a Lie-image of M, we mean the object in S^N corresponding to the Lie-image of the figure in Q obtained from M.

Above explanation is not explicite enough, but for the moment, imagine as a Lie-image of M, for example, an image of M obtained by repeating the following operation several times: take a parallel hypersurface of a conformal image of M.

Note that a Lie-image of a regular hypersurface is not necessarily regular. But generally, even if we consider only Lie-transformations with regular images, we have $\{\text{Lie-images of } M\} \supsetneqq \{\text{conformal images of } M\}$. The following facts reveal the reason why before Pinkall, no one considered Lie-transformations.

Fact 2.1 $\{\text{Regular Lie-images of } S^n(r)\} = \{\text{conformal images of } S^n(r)\}$.

Fact 2.2 {Regular Lie-images of $S^k(r_0)\times S^{n-k}(\sqrt{1-r_0^2})\} \subset \bigcup_r\{$conformal images of $S^k(r)\times S^{n-k}(\sqrt{1-r^2})\}$ = {Cyclides of Dupin in $S^{n+1}(1)$}.

§3. Proof of Theorem I.

Let $f: M^n \longrightarrow E^{n+1}$ be an immersed hypersurface in E^{n+1} with a global unit normal vector field ξ and let A be the second fundamental form of f. By $\lambda_1 \leqq \cdots \leqq \lambda_n$, we denote the continuous principal curvature functions on M. For $p \in M$ and $\lambda \in \{\lambda_1,\ldots,\lambda_n\}$, let $E_p(\lambda) = \{X \in T_pM \mid AX = \lambda X\}$. Then the following facts are fundamental [R]:

Fact 3.1

(a) The function $M \ni p \longmapsto \dim E_p(\lambda)$ is upper-semicontinuous, i.e. the subset $\{p \in M \mid \dim E_p(\lambda) \leqq k\}$ is open in M for $1 \leqq k \leqq n$.

(b) Let G_k be the interior of the subset $\{p \in M \mid \dim E_p(\lambda) = k\}$, $1 \leqq k \leqq n$. Then $\bigcup_k G_k$ is open and dense in M.

(c) On G_k, the vector spaces $E_p(\lambda)$, $p \in G_k$, form an integrable distribution E_k.

(d) If $k \geqq 2$, every leaf of E_k on G_k is a part of k-dimensional sphere of E^{n+1}.

(e) When M is complete, let $k_0 = \min\{k \mid G_k \neq \phi\}$. Then every leaf of E_{k_0} on G_{k_0} is complete.

Thus, for the proof of our theorem, it is sufficient to show that a similar fact holds for k = 1 in (d) above, in case M is a tautly embedded compact hypersurface. We prove this in a stronger version as follows:

Proposition 3.2 Let M be a tautly embedded compact hypersurface in E^{n+1}. For some principal curvature function λ of M, let $G_1 \neq \phi$. Then every leaf of E_1 on G_1 is a closed circle in E^{n+1}.

Proof. The part "closed" follows from (e) above.

Let X be a principal vector field on G_1 with respect to λ. It is sufficient to prove $X(\lambda) \equiv 0$ on G_1 (cf. [CR0]). To do this, we suppose $\lambda<0$ and $X(\lambda)>0$ at a point $p \in G_1$. Then in a neighbourhood U of p in G_1, we have $\lambda<0$ and $X(\lambda)>0$. Let $\gamma(t)$ be the leaf of E_1 through $p = \gamma(0)$. We denote by $B(x,r)$ the closed ball in E^{n+1} with center at $x \in E^{n+1}$ and radius $r>0$. Let $f_p^\lambda = p + \lambda(p)^{-1}\xi_p$ and let m be the sum of the multiplicities of focal points of p lying on the closed segment $\overline{pf_p^\lambda}$.

Claim. For $q = \gamma(t_1) \in U$, $t_1>0$, we have $\bar\lambda(p) + |f_p^\lambda - f_q^\lambda| < \bar\lambda(q)$, where we put $\bar\lambda = -\lambda^{-1}$.

Consider a curve $\alpha(t)$ connecting f_p^λ and f_q^λ, given by

$$\alpha(t) = \gamma(t) + \lambda(\gamma(t))^{-1}\xi_{\gamma(t)}.$$

Noting that α dose not lie on a straight line and

$$\dot{\alpha}(t) = -X(\lambda(\gamma(t)))\lambda(\gamma(t))^{-2}\xi_{\gamma(t)},$$

we get

$$|f_p^\lambda - f_q^\lambda| < \int_0^{t_1}|\dot{\alpha}(t)|\,dt = [-\lambda(\gamma(t))^{-1}]_0^{t_1} = \bar{\lambda}(q) - \bar{\lambda}(p).$$

This claim is due to [CR1], from which we get

$$B(f_p^\lambda,\bar{\lambda}(p)) \subsetneqq B(f_q^\lambda,\bar{\lambda}(q)).$$

Now, for some $\varepsilon>0$, let $x_q^\varepsilon = q+(\bar{\lambda}(q)+\varepsilon)(-\xi_q) = f_q^\lambda - \varepsilon\xi_q$. We may assume that $L_{x_q^\varepsilon}$ is a Morse function on M. (Otherwise, choose $y \in E^{n+1}$ very near to x_q^ε such that L_y is a Morse function and has a critical point $\tilde{q}$ very near to q with index m [CW, Lemma(3.1)]. The following argument holds for y and $\tilde{q}$ instead of x_q^ε and q with a slight modification).

Put $r_q^\varepsilon = \min\{r \mid B(x_q^\varepsilon,r) \supset B(f_p^\lambda,\bar{\lambda}(p)\}$. Then it is easy to see that $r_q^\varepsilon = \bar{\lambda}(p) + |f_p^\lambda - x_q^\varepsilon|$. We have $r_q^\varepsilon < \bar{\lambda}(q)+\varepsilon$ since

$$\bar{\lambda}(q)+\varepsilon>\bar{\lambda}(p)+|f_p^\lambda-f_q^\lambda|+\varepsilon = \bar{\lambda}(p)+|f_p^\lambda-f_q^\lambda|+|f_q^\lambda-x_q^\varepsilon| \geq \bar{\lambda}(p)+|f_p^\lambda-x_q^\varepsilon|.$$

Now, take $\delta>0$ such that $\bar{\lambda}(q)+\varepsilon-r_p^\varepsilon>\delta$. Then $B(x_q^\varepsilon,r_q^\varepsilon+\delta) \supsetneqq B(f_p^\lambda,\bar{\lambda}(p))$ and $B(x_q^\varepsilon,r_q^\varepsilon+\delta) \not\ni q$. Now, from tautness, for $\delta_0>0$,

$$\beta_m(B(f_p^\lambda,\bar{\lambda}(p)+\delta_0)\cap M,\mathbb{Z}_2) \geq 1.$$

Since $B(x_q^\varepsilon,r_q^\varepsilon+\delta) \supset B(f_p^\lambda,\bar{\lambda}(p)+\delta_0)$ for sufficiently small δ_0, it follows again from tautness that $L_{x_q^\varepsilon}$ has a critical point with index m in $B(x_q^\varepsilon,r_q^\varepsilon+\delta)$. On the other hand, since $q \notin B(x_q^\varepsilon,r_q^\varepsilon+\delta)$ is also a critical point of $L_{x_q^\varepsilon}$ with index m, we have

$$\beta_m(B(x_q^\varepsilon,\bar{\lambda}(q)+\varepsilon)\cap M,\mathbb{Z}_2) \geq 2.$$

Applying this argument to $q' = \gamma(t_2) \in U$, $t_1<t_2$, we get for some $\varepsilon'>0$,

$$\beta_m(B(x_{q'}^{\varepsilon'},\bar{\lambda}(q')+\varepsilon')\cap M,\mathbb{Z}_2) \geq 3.$$

Now, repeating this argument along $\{\gamma(t)\} \subset U$, we get finally $\beta_m(M,\mathbb{Z}_2) \longrightarrow \infty$, a contradiction. Thus we get $X(\lambda) \equiv 0$ on G_1. Q.E.D.

§4. Taut embeddings $M^3 \subset \bar{M}^4$.

In the forth-coming paper, we will give a complete proof of Theorem II in the following way.

When the number h of the pricipal curvatures is constant on M, i.e.

when <u>D1</u> is satisfied on M, M is a Dupin hypersurface by Theorem I. Since compact embedded Dupin hypersurfaces with h = 1,2,3 are classified already (cf. §0), the problem remains when h is not constant on M. We know that M has no umbilics unless $M = S^3(r)$ ([CW]). So we assume $M = M_2 \cup M_3$ where $M_i = \{p \in M | h(p) = i\}$. By Proposition 3.2, M_3 is foliated by three families of circles. This fact is extended to any boundary point $p \in \partial M_3 \subset M_2$. Investigating the situation around p, we get the fourth case in the classification.

§5. Open problems.

The next step to classify taut embeddings would begin in solving the following questions.

<u>Q1</u>. Classify compact embedded Dupin hypersurfaces with h = 4 or 6. Are they Lie-images of isoparametric hypersurfaces?

<u>Q2</u>. Classify taut embeddings $M^n \subset \bar{M}^{n+1}$ such that $\dim H^*(M, \mathbb{Z}_2) = 6$, 8,12. Then we know all taut embeddings $M^n \subset M^{n+1}$ such that the multiplicities of the smallest and largest principal curvatures are constant on M.

The next problem occurs when the multiplicity of the smallest or the largest principal curvature changes. At the changing point $p \in M$, the following problem is very difficult, but very interesting:

<u>Q3</u>. Let f_p be the first focal point of p in a normal direction ξ and let $r = d(p, f_p)$. From tautness, $B(f_p, r) \cap M \subset \partial B(f_p, r) =: S_p$.

(a) Is $B(f_p, r) \cap M$ a differential submanifold of S_p?

(b) Is $B(f_p, r) \cap M$ taut in S_p?

The answer seems to be affirmative for an analytic manifold M.

REFERENCES

[B] Banchoff, T.F., The spherical two-piece property and tight surfaces in spheres, J.Diff.Geom. 4(1970) 193-205.

[CW] Carter, S. and West, A., Tight and taut immersions, Proc. London Math. Soc. 25(1972) 701-720.

[CR0] Cecil, T.E. and Ryan, P.J., Focal sets of submanifolds, Pacific J.Math. 78(1978) 27-39.

[CR1] ______________________, Focal sets, taut embeddings and the cyclides of Dupin, Math.Ann. 236(1978) 177-190.

[CR2] ______________________, Tight spherical embeddings, Lecture Note in Math., Springer 838(1981) 94-104.

[CL] Chern, S.S. and Lashof, R.K., On the total curvature of immersed manifolds, Amer.J.Math. 79(1957) 306-318.

[K] Kuiper,N.H., On convex maps, Nieuw Archief voor Wiskunde, 10(1962) 147-164.

[M] Miyaoka,R., Compact Dupin hypersurfaces with three principal curvatures, to appear in Math. Zeit.

[NR] Nomizu,K. and Rodriguez,L., Umbilical submanifolds and Morse functions, Nagoya Math.J. 48(1972) 197-201.

[OT] Ozeki,H. and Takeuchi,M., On some types of isoparametric hypersurfaces in spheres I,II, Tohoku Math. J. 27(1975) 515-559, 28 (1976) 7-55.

[P] Pinkall,U., Dupin'sche Hyperflächen, Thesis, Freiburg i.Br.,(1981)

[R] Reckziegel,H., Completeness of curvature surfaces of an isometric immersion, J.Diff.Geom. 14(1979) 7-20.

[TK] Takeuchi,M. and Kobayashi,S., Minimal imbeddings of R-spaces, J. Diff.Geom. 2(1968) 203-215.

[T] Thorbergsson,G., Dupin hypersurfaces, Bull. London Math. Soc. 15 (1983) 493-498.

GEOMETRIC BOUNDS FOR THE NUMBER OF CERTAIN HARMONIC MAPPINGS

T. Adachi and T. Sunada*

Department of Mathematics
Nagoya University
Nagoya 464/ Japan

1. We will start with giving some motives of this investigation which we borrow from geometric analogues of a few number-theoretic theorems. The first one is the celebrated prime number theorem, which asserts that the number of prime numbers not exceeding x is asymptotically equal to $x/\log x$ as x tends to infinity. The second is the Mordell conjecture (established recently by Faltings) on finiteness of rational points on a curve with genus greater than one, defined over the rational numbers $\mathbb{Q}$.

To give a geometric analogue of the first, we let N be a compact Riemannian manifold with negative sectional curvature, and P the set of all prime geodesic cycles in N. The following shows that prime geodesic cycles play a similar role as prime numbers.

Theorem 1 (A special case of Parry and Pollicott [23]) *Let h_N denotes the exponential growth rate of volume of the geodesic balls in the universal covering of* N:

$$h_N = \lim_{R\to\infty} R^{-1}\log(\mathrm{Vol}(B_R(x)), \quad B_R(x) = \{y \,;\, d(y, x) \leq R\}.$$

Then

$$\#\{p \in P \,;\, \exp(h_N(\text{length of } p)) \leq x\} \sim x/\log x.$$

In particular, one gets

---

*Supported by The Ishida Foundation.

$$\lim_{x\to\infty} x^{-1}\log \#\{p \in P \ ; \ (\text{length of } p) \leq x\} = h_N.$$

For the second, we first replace the rationals $\mathbb{Q}$ by a function field K defined over the complex numbers $\mathbb{C}$, and consider a smooth projective algebraic variety $\mathfrak{X}$ defined over K instead of a curve. We assume a condition, corresponding to the genus condition of curves, that the tangent bundle of $\mathfrak{X}$ is negative (or equivalently the cotangent bundle is ample in the algebraic-geometric sense). Then either

i) the K-rational points $\mathfrak{X}(K)$ of $\mathfrak{X}$ lies in a proper algebraic subset of $\mathfrak{X}$, or

ii) $\mathfrak{X}$ is K-isomorphic to a smooth projective algebraic variety $\mathfrak{X}_0$ defined over $\mathbb{C}$, and $\mathfrak{X}_0(K) - \mathfrak{X}_0(C)$ is finite.

(see Noguchi [21] for the proof).

If K is a function field on a smooth algebraic variety M defined over $\mathbb{C}$, then a K-variety $\mathfrak{X}$ yields a fiber space $X \longrightarrow M$ whose generic fiber N has negative tangent. The set $\mathfrak{X}(K)$ is the space of rational sections of the fiber space. In this situation, the second case in the above reduces to

Theorem 2 (A special case of Noguchi and Sunada [22]). *If M is a compact algebraic variety, and N is a smooth projective variety with negative tangent bundle, then the set of non-constant holomorphic mappings of M into N is finite.*

2. We will try to give a geometric unification to the examples in the previous section. Let M and N be compact connected Riemannian manifolds. For a smooth mapping φ of M into N, we define the energy integral $E(\varphi)$ to be

$$\int_M \|d\varphi\|^2 \, dv_M .$$

The critical points of the functional E are called harmonic mappings. We denote by $H(M, N)$ the set of all harmonic mappings of M into N. In the case M is the circle $\mathbb{R}/\mathbb{Z}$, harmonic mappings are just closed geodesics, and for any closed geodesic c, we find $E(c) = (\text{length of } c)^2$. If M and N are Kähler manifolds, then holomorphic mappings of M into N are harmonic. Bearing these examples in mind, it is natural to generalize the results in §1 in terms of harmonic mappings .

From now on, we assume that N is non-positively curved. In a loose sense, this corresponds to the negativity of tangent bundle. The facts which we freely make use of are

1) Each component of the space of continuous mappings contains one and only one component of $H(M, N)$.

2) On each component H_α of $H(M, N)$, the functional E assumes a constant value, which we denote by $E(H_\alpha)$.

One of main results in this note is

Theorem A. *There are positive constants c_1 and c_2 depending only upon the diameters D_M, D_N, the volumes V_M, V_N and the lower limits of the Ricci curvature of M and N such that for any x*

$$\#\left\{ H_\alpha \subset H(M, N)\ ;\ E(H_\alpha) \leqq x^2 \right\} \leqq c_1 \exp(c_2 x).$$

The constants c_1 and c_2 can be explicitly calculated in terms of D, V and the Ricci curvatures.

Remark. In general case, not much is known on the structure of $H(M, N)$, and the aspect of the energy spectrum $\{E(\varphi)\ ;\ \varphi \in H(M, N)\}$ might be possibly complicated because we may not expect the Palais-Smale condition for the functional E. But for manifolds N with non-positive curvature, a weak compactness property holds, if not the P-S condition, which allows us to prove discreteness of the energy spectrum as above.

The following which is a byproduct of the proof of Theorem A is a partial generalization of Theorem 1.

Theorem B. i) There exists a constant c_3 depending only upon D_M, V_M and the lower limit of the Ricci curvature of M such that

$$\limsup_{x \to \infty} x^{-1} \log \# \{ H_\alpha \subset H(M, N) \; ; \; E(H_\alpha) \leqq x^2 \} \leqq c_3 h_N .$$

ii) Let $T(M, N)$ be the space of totally geodesic mappings of M into N. Suppose that the fundamental group of M is generated by homotopy classes of geodesic loops $\gamma_1, \ldots, \gamma_t$ with length $\gamma_i \leqq \ell$. Then

$$\limsup_{x \to \infty} x^{-1} \log \# \{ H_\alpha \subset T(M, N) \; ; \; E(H_\alpha) \leqq x^2 \} \leqq \ell t \, V_M^{-1/2} h_N .$$

Especially

$$\limsup_{x \to \infty} x^{-1} \log \# \{ H_\alpha \subset H(\mathbb{R}^m/\mathbb{Z}^m, N) \; ; \; E(H_\alpha) \leqq x^2 \} \leqq m h_N .$$

In the above, it should be noted that if a component $H_\alpha \subset H(M, N)$ contains a totally geodesic mapping, then all the mappings in H_α are also totally geodesic, and that if the Ricci curvature of M is semi-positive definite, then any harmonic mappings are totally geodesic.

3. Let M and N be Kähler manifolds. As a corollary of Theorem A we can deduce some finiteness theorems for holomorphic mappings. To see this we first make

Definition. A pair of complete Kähler manifolds (M, N) is called to satisfy the Schwartz-Ahlfors property if there exists a positive constant K such that $\| d\varphi \| \leqq K$ for any holomorphic mappings φ of M into N.

The classical Schwartz lemma implies that the pair of the unit discs with the Poincare metrices satisfies the S-A property.

Proposition 1. Let M and N be compact Kähler manifolds. Suppose

<u>that the sectional curvature of</u> N <u>is non-positive</u>. <u>Further if the pair</u> (M, N) <u>satisfies the</u> S-A <u>property with a constant</u> K, <u>then there exists only finitely many components of</u> Hol(M, N), <u>the space of holomorphic mappings</u>, <u>and the number of components is estimated in terms of the constants</u> c_1, c_2 <u>and</u> K.

This is an easy consequence of the fact that the components of Hol(M, N) are also components of H(M, N) and $E(\varphi) \leq K^2 V_M$ for any $\varphi \in$ Hol(M, N). We should point out here (Royden[24]) that if the holomorphic sectional curvature of N is bounded from above by a negative constant, then for any M, the pair (M, N) satisfies the S-A property.

We now turn to general situation. We set, for a subspace V in T_yN,

$$V^{(0)} = \{ w \in T_yN \ ; \ \langle R(w, v)w, v \rangle = 0 \ \text{ for any } \ v \ \text{ in } \ V \}.$$

Non-positivity of the sectinal curvature of N guarantees that $V^{(0)}$ is also a subspace. We define numbers n(y) and n(N) by

$$n(y) = \max \{ \dim V \ ; \ V^{(0)} \neq 0 \}$$

$$n(N) = \max \{ n(y) \ ; \ y \in N \}.$$

If N is negatively curved at $y \in N$, then $n(y) = 1$, and if the Ricci curvature is negative definite at y, then $n(y) \leq n - 1$.

<u>Proposition 2</u>. <u>If for some</u> x <u>in</u> M, $\operatorname{rank} d_x\varphi > n(\varphi(x))$, <u>then</u> φ <u>is rigid in the sense that there is no other harmonic mapping homotopic to</u> φ. <u>In other words</u>, <u>the component of</u> H(M, N) <u>containing</u> φ <u>is the singleton</u> $\{\varphi\}$. <u>In particular</u>, <u>if the Ricci curvature of</u> N <u>is negative at some point</u>, <u>and if</u> $\varphi : M \longrightarrow N$ <u>is a surjective harmonic mapping</u>, <u>then</u> φ <u>is rigid</u>.

An out-line of proof is the following: For homotopic $\varphi, \psi \in$ H(M, N), we take a unique section $X \in C^{\infty}(\varphi^{-1}TN)$ satisfying $\psi(x) = \operatorname{Exp}_{\varphi(x)} X(x)$. Fixing an orthonormal basis $\{e_1, \ldots, e_m\}$ in T_xM, we define a family of surfaces $\alpha_j : (-\varepsilon, \varepsilon) \times [0, 1] \longrightarrow N$ by setting

$$\alpha_j(s, t) = \operatorname{Exp} tX(\operatorname{Exp} se_j), \quad j = 1, \ldots, m.$$

Then the second variation formula leads to

$$\frac{1}{2}(\Delta \|X\|^2)(x) = \sum_{j=1}^{m} \int_0^1 \left\{ \left\| \frac{D}{\partial t} \frac{\partial \alpha_j}{\partial s} \right\|^2 - \left\langle R\left(\frac{\partial \alpha_j}{\partial s}, \frac{\partial \alpha_j}{\partial t}\right) \frac{\partial \alpha_j}{\partial s}, \frac{\partial \alpha_j}{\partial t} \right\rangle \right\}_{(0,t)} dt.$$

Note here

$$\frac{D}{\partial t} \frac{\partial \alpha_j}{\partial t} = o, \quad \frac{D \alpha_j}{\partial t}\bigg|_{(0,0)} = X(x), \quad \frac{\partial \alpha_j}{\partial t}\bigg|_{(0,1)} = \frac{d}{dt}\bigg|_{t=1} \operatorname{Exp} tX(x)$$

$$\sum_{j=1}^{m} \frac{D}{\partial s} \frac{\partial \alpha_j}{\partial s}\bigg|_{(0,0)} = \operatorname{tr} \nabla d \varphi = o,$$

$$\sum_{j=1}^{m} \frac{D}{\partial s} \frac{\partial \alpha_j}{\partial s}\bigg|_{(0,1)} = \operatorname{tr} \nabla d \psi = o.$$

from which it follows that $\Delta \|X\|^2 \geqq 0$, so that $\|X\|$ is constant, and

$$\left\| \frac{D}{\partial t} \frac{\partial \alpha_j}{\partial s} \right\| = 0, \quad \langle R(d_x \varphi(e_j), X(x)) d_x \varphi(e_j), X(x) \rangle = 0.$$

From the assumption, there exists a point x in M such that

$$\dim(\operatorname{Im}(d_x \varphi)) > n(\varphi(x)),$$

whence $o = X(x) \in (\operatorname{Im} d_x \varphi)^{(0)}$, or $X(x) = o$. This implies $\varphi \equiv \psi$.

Remark. In case N is locally symmetric, $n(N)$ can be explicitly calculated. Moreover if N is a compact quotient of a symmetric bounded domain D, $n(N)$ is closely related to the proper boundary components of D.

Corollary. *If N has negative holomorphic sectional curvature and non-positive sectional curvatutre, then*

$$\# \{ \varphi \ \mathrm{Hol}(M, N) \ ; \ \operatorname{rank} \varphi > n(N) \}$$

is finite, and estimated as in Proposition 1.

As for finiteness of holomorphic mappings, we have a satisfactory generalization: Let M and N be compact projective algebraic varieties. We denote by $R^k(M, N)$ the set of rational mappings φ of M into N with $\dim \varphi(M) \geq k$.

Theorem 3. *If the k-th exterior product* $\Lambda^k TN$ *is negative, then* $\# R^k(M, N)$ *is finite*.

See Noguchi and Sunada [22] for the proof. The case $k=1$ is Theorem 2.

Another consequence of Proposition 2 is

Proposition 3. *If* N *is non-positively curved and there exists a point* $y \in N$ *such that the Ricci curvature is negative definite at* y, *then the order of the isometry group of* N *is estimated in terms of* V_N, D_N *and the lower limit of the Ricci curvature of* N.

Remark. i) If we allow to include the injectivity radius in the estimation of the order, then we may much more easily get a bound of the order in the following type:

$$\# \mathrm{Iso}(N) \leq k^k, \qquad k = \frac{n}{\mathrm{Vol}(S^{n-1}(1))} V_N \left(\frac{8}{i_N}\right)^n ,$$

Recently, A. Katsuda succeeded in giving a geometric bound for the order of the isometry group of a manifold with negative Ricci curvature without the assumption of non-positivity of sectional curvature. A generalization for non-compact case has been given by Yamaguchi [31].

ii) Proposition 3 is regarded as a partial generalization of the classical theorem by Hurwitz that, for a compact Riemann surface N with constant negative curvature -1, $\# \mathrm{Iso}(N) \leq 42\, V_N / \pi$.

4. We shall give an outline of a proof of Theorem A. We denote by A_x the set of components of $H(M, N)$ containing some φ such that $\sup \| d\varphi \| \leq x$. It is known that if φ and ψ are homotopic, then $\| d\varphi \| = \| d\psi \|$. The Sobolev embedding theorem in the line of Uhlenbeck [30] means that it is enough to estimate $\# A_x$. In fact

Lemma. There is a computable constant c depending only on D_M, V_M and the lower limit of the Ricci curvature of M such that for any $\varphi \in H(M, N)$

$$\sup \|d\varphi\| \leq cE(\varphi).$$

Since N is a $K(\pi, 1)$-space, homotopy classes of mappings of M into N are parametrized by conjugacy classes of induced homomorphisms : $\pi_1(M, x_0) \longrightarrow \pi_1(N, y_0)$. We choose generators of $\pi_1(M)$

$$T = \{\sigma_1, \ldots, \sigma_t\},$$

which are represented by geodesic loops of length $\leq 2D_M + \varepsilon$. We let φ be a harmonic mapping lying in a component in A_x. Then $\varphi_*(\sigma_i)$ is an element in $\pi_1(N)$ represented by a geodesic loop of length $\leq (2D_M + \varepsilon)x + 2D_N$. Thus if we set

$$S_x = \{\gamma \in \pi_1(N) \; ; \; d_{\tilde{N}}(\tilde{y}_0, \gamma\tilde{y}_0) \leq (2D_M + \varepsilon)x + 2D_N\},$$

d_N being the distance function on the universal covering $\tilde{N}$, then $\#A_x \leq (\#S_x)^{\#T}$. Thus we are led to the problem of counting lattice points (an orbit of the deck transformation group) in the geodesic balls in the universal covering.

Lemma. Let X^k be a compact connected Riemannian manifold of k-dimension, and let x_0 be a point in the universal covering $\tilde{X}$. Then for any positive R,

$$\#\{\gamma \in \pi_1(X) \; ; \; d_{\tilde{X}}(x_0, \gamma x_0) \leq R\} \leq \frac{\mathrm{Vol}(S^{k-1}(1)) \exp((k-1)|\rho_X|^{1/2}(R+D_X))}{(k-1)V_X \, |\rho_X|^{k/2} \, 2^{k-1}},$$

where $\rho_X = \inf \mathrm{Ricci}(v)/(k-1)$, $v \in TX$, $\|v\| = 1$.

Proof. Let $\mathcal{D}$ be a fundamental domain in X such that $x_0 \in \mathcal{D} \subset B_{D_X}(x_0)$. If γ runs over elements in $\pi_1(X)$ with $d_{\tilde{X}}(x_0, \gamma x_0) \leq R$, then

$$\bigcup_{\gamma} \gamma \mathcal{D} \subset B_{R+D_X}(x_0),$$

so that we have

$$\#\{\gamma \ ; \ d_{\tilde{X}}(x_0, \gamma x_0) \leq R\} \leq V_X^{-1} \operatorname{Vol}(B_{R+D_X}(x_0)).$$

On the other hand, the volume comparison theorem says

$$\operatorname{Vol}(B_r(x_0)) \leq \operatorname{Vol}(S^{k-1}) \int_0^r \left(\frac{\sinh |\rho_X|^{1/2} t}{|\rho_X|^{1/2}} \right)^{k-1} dt,$$

whence the lemma.

Applying this lemma to the cases $X = M$, $R = 2(D_M + \varepsilon)$ and $X=N$, $R = 2x(D_M + \varepsilon) + 2D_N$, we have

$$\#A_x \leq c_1(D_M, V_M, \rho_M, D_N, V_N, \rho_N) \exp(c_2(D_M, V_M, \rho_M) |\rho_N|^{1/2} x).$$

This completes the proof.

The proof of i) in Theorem B is the following: In view of the above argument, we find

$$\#\{H_\alpha \subset H(M, N) \ ; \ E(H_\alpha) \leq x^2\} \leq \{V_N^{-1} \operatorname{Vol}(B_{2\sqrt{c}D_M x + 3D_N}(y_0))\}^{\#T}$$

so that

$$\begin{aligned} &\limsup_{x\to\infty} x^{-1} \log \#\{H_\alpha \ ; \ E(H_\alpha) \leq x^2\} \\ &\leq \#T \limsup_{x} x^{-1} \log \operatorname{Vol}(B_{2\sqrt{c}D_M x}(y_0)) \\ &= \#T 2\sqrt{c} D_M \limsup_{x} x^{-1} \log \operatorname{Vol}(B_x(y_0)) \\ &= c_3 h_N. \end{aligned}$$

For the detail, see [1].

5. We conclude with proposing several open problems which are closely related to existence problems of harmonic mappings.

i) Let M and N be compact Riemann surfaces with negative sectional curvature. Then is the following positive?

$$\liminf_{x\to\infty} x^{-1}\log \#\{H_\alpha \subset H(M,N);\ \exists\varphi\in H_\alpha \text{ with } \operatorname{rank}\varphi = 2, \text{ and } E(H_\alpha) \leqq x^2\}.$$

ii) Let N be a locally symmetric space of non-positive curvature whose rank is r, then is

$$\liminf_{x\to\infty} x^{-1}\log\#\{H_\alpha\subset H(\mathbb{R}^r/\mathbb{Z}^r,N);\ \exists\varphi\in H_\alpha \text{ with } \operatorname{rank}\varphi = r,\ E(H_\alpha)\leqq x^2\}$$

positive?

iii) Let N be a compact Riemannian manifold with non-positive curvature. Then is it true that

$$\lim_{x\to\infty} x^{-1}\log \#\{H_\alpha\subset H(\mathbb{R}/\mathbb{Z},\ N)\ ;\ E(H_\alpha)\leqq x^2\} = h_N\ ?$$

iv) Let N be non-positively curved, and let $\rho : H_1(M, \mathbb{Z}) \longrightarrow H_1(N, \mathbb{Z})$ be a homomorphism. Then does there exist the limit

$$\lim_{x\to\infty} \frac{\#\{H_\alpha\subset H(M,N);\ \exists\varphi\in H_\alpha \text{ with } \varphi_* = \rho,\ \text{and } E(H_\alpha)\leqq x^2\}}{\#\{H_\alpha\subset H(M,N)\ ;\ E(H_\alpha)\leqq x^2\}}\ ?$$

References

[1] T. Adachi and T. Sunada; Energy spectrum of certain harmonic mappings, preprint, 1983.

[2] R.L. Bishop and R.J. Crittenden; Geometry of Manifolds, Academic Press: New York and London 1964.

[3] P. Buser and H. Karcher; Gromov's almost flat manifolds, Asterisque 81(1981).

[4] J. Eells and L. Lemaire; A report on harmonic maps, Bull. London Math. Soc. 10(1978), 1-68.

[5] J. Eells and J. Sampson; Harmonic mappings of Riemannian manifolds, Am. J. Math. 86(1964).

[6] S. Gallot; Inégalité isoperimetrique sur les variétés compacte sans bord, to appear.

[7] R. Gangolli; On the length spectra of certain compact manifolds of negative curvature, J. Diff. Geom. 12(1977), 403-424.

[8] M. Gromov; Structures metriques sur les variétés riemanniennes (rédigé par J. Lafontaine et P. Pansu), Cedic: Paris, 1982.

[9] P. Hartman; On homotopic harmonic maps, Can. J. Math. 19(1967), 673-687.

[10] H.Huber; Über die Isometriegruppe einer kompakten Mannigfaltigkeit mit negativer Krümmung, Helv. Phys. Acta 45(1972), 277-288.

[11] H.C. Im Hof; Über die Isometriegruppe bei kompakten Mannigfaltig keiten negativer Krummung, Comment. Math. Helv. 48(1973), 14-30.

[12] M. Kalka, B. Shiffman and B. Wong; Finiteness and rigidities theorem for holomorphic mappings, Michigan Math. J. 28(1981), 289-295.

[13] A. Katsuda; The isometry groups of compact manifolds with negative Ricci curvature, preprint, 1983.

[14] G. Knieper; Das Wachstum der Aquivalenzklassen geschlossener Geodätischer in kompakten Mannigfaltikeiten, Arch. Math. 40(1983),

559-568.

[15] S. Kobayashi; Hyperbolic Manifolds and Holomorphic Mappings, Marcel Dekker, New York, 1970.

[16] L. Lemaire; Harmonic mappings of uniform bounded dilatation, Topology, 16(1977), 199-201.

[17] P. Li; On the Sobolev constant and the p-spectrum of a compact Riemannian manifolds, Ann. scient. Ec. Norm. Sup. 13(1980), 451-469.

[18] A. Lichnerowicz; Applications harmoniques et variétés kahlériennes Symp. Math. III, Bologna, (1970), 341-402.

[19] M. Maeda; The isometry groups of compact manifolds with non-positive curvature, Pro. Japan Acad. 51(1975), 790-794.

[20] A. Mannings; Topological entropy for geodesic flows, Ann. of Math. 110(1979), 567-573.

[21] J. Noguchi; A higher dimensional analogue of Mordell's conjecture over function fields, Math. Ann. 258(1981), 207-212.

[22] J. Noguchi and T. Sunada; Finiteness of the family of rational and meromorphic mappings into algebraic varietes, Am. J. Math. 104 (1982), 887-900.

[23] W. Parry and M. Pollicott; An analogue of the prime number theorem for closed orbits of Axiom A flows, Ann. of Math. 118(1983), 573-591.

[24] H.L. Royden; The Ahlfors-Schwarz lemma in several complex variables, Comment. Math. Helv. 55(1980), 547-558.

[25] R. Schoen and S.T. Yau; Compact group actions and the topology of manifolds with non-positive curvature, Topology 18(1979), 361-380.

[26] T. Sunada; Holomorphic mappings into a compact quotient of symmetric bounded domain, Nagoya Math. J. 64(1976), 159-175.

[27] T. Sunada; Rigidity of certain harmonic mappings, Invent. Math. 51(1979), 297-307.

[28] T. Sunada; Tchebotarev's density theorem for closed geodesics in a compact locally symmetric space of negative curvature, preprint.

[29] T. Sunada; Geodesic flows and geodesic random walks, Advanced Studies in Pure Math. 3(1984), Geometry of Geodesics and Related Topics, 47-85.

[30] K. Uhlenbeck; Morse theory by perturbation methods with applications to harmonic maps, Trans. AMS. 267(1981), 569-583.

[31] T. Yamaguchi; On the isometries of negatively curved manifolds with finite volume, preprint, Univ. of Tsukuba, 1983.

[32] A. Howard and A.J. Sommese, On the orders of the automorphism groups of certain projective manifolds, Progress in Math. 14 (1981), 145-158.

THE FIRST STANDARD MINIMAL IMMERSIONS OF COMPACT IRREDUCIBLE SYMMETRIC SPACES

Yoshihiro Ohnita
Mathematical Institute
Tohoku University
Sendai 980
Japan

Introduction. Let $M = G/K$ be an n-dimensional irreducible symmetric space of compact type and g_0 be the G-invariant Riemannian metric induced by the Killing form of the Lie algebra of G. Let Δ be the Laplace-Beltrami operator of (M,g_0) acting on the space of all real-valued C^∞-functions on M. For the k-th eigenvalue λ_k of Δ we can construct a minimal isometric immersion ϕ_k of $(M,(\lambda_k/n)g_0)$ into $S^{m(k)}(1)$ using an orthonormal basis of the k-th eigenspace V_k where $m(k)+1 = \dim V_k$ (cf. Takahashi [11]). We call ϕ_k the k-th standard minimal immersion of M. We call that two isometric immersions ϕ,ψ of (M,g) into $S^m(1)$ are equivalent if there exists an isometry ρ of $S^m(1)$ such that $\psi = \rho\cdot\phi$ and that a minimal isometric immersion ϕ is rigid if every minimal isometric immersion ψ of (M,g) into $S^m(1)$ is equivalent to ϕ. In this paper we study the rigidity of the first standard minimal immersions of compact irreducible symmetric spaces.

Now let M be a sphere or projective space, that is, one of the following : S^n, $P_n(\mathbb{R})$, $P_n(\mathbb{C})$, $P_n(\mathbb{H})$ and $P_2(\mathbb{C}ay)$. They are the most fundamental compact irreducible symmetric spaces. The first standard minimal immersions of projective spaces are what we call the generalized Veronese submanifolds (cf. Sakamoto [10]).

Theorem. (do Carmo and Wallach [1], Mashimo [5])
(1) If M is a sphere, then the second standard minimal immersion ϕ_2 is rigid.
(2) If M is a projective space, then the first standard minimal immersion ϕ_1 is rigid.

Remark. The first standard minimal immersion of a sphere is the identity. The composition of the isometric covering $S^n \to P_n(\mathbb{R})$ and the first standard minimal immersion of $P_n(\mathbb{R})$ is the second standard minimal immersion of S^n.

From this theorem we pose the following problem :

Problem. For any compact irreducible symmetric space M , is the first standard minimal immersion ϕ_1 of M rigid ?

The purpose of this paper is to study this problem. Our main results are the following theorems.

Theorem A. Let $M = G/K$ be one of the following spaces:
(1) simply connected compact irreducible symmetric spaces of classical type except $SO(9)/SO(6)\times SO(3)$ and $SO(8)/SO(4)\times SO(4)$,
(2) non simply connected spaces $SO(p+q)/S(O(p)\times O(q))$ $(p\geqq q>1)$, $Sp(4)/Sp(4)\times Sp(4)\cdot\mathbb{Z}_2$ and $SU(8)/Sp(4)\cdot\mathbb{Z}_2$,
(3) simply connected exceptional spaces $E_6/Spin(10)\cdot T^1$, $E_7/E_6\cdot T^1$ and E_6/F_4 .
Then the first standard minimal immersion of M is rigid.

Theorem B. Let $M = G/K$ be one of $SO(9)/SO(6)\times SO(3)$ and $SO(8)/SO(4)\times SO(4)$. Then the first standard minimal immersion of M is nonrigid.

1. The standard minimal immersions

In this section we explane the costruction of the standard minimal immersions and their properties (cf. Wallach [16]).

Let $M = G/K$ be an n-dimensional irreducible symmetric space of compact type. (G,K) is a symmetric pair and G is a compact semisimple Lie group. We denote by $\mathfrak{g}$ and $\mathfrak{k}$ the Lie algebra of G and K, respectively. Let g_0 be the G-invariant Riemannian metric on M induced by (-1)-times the Killing form B of $\mathfrak{g}$. Let Δ be the Laplace-Beltrami operator of (M,g_0) acting on the space $C^\infty(M)$ of all real-valued C^∞-functions on M. We denote by $0 = \lambda_0 < \lambda_1 < \dots < \lambda_k < \dots$, the set of all mutually distinct eigenvalues of Δ, and by V_k the eigenspace of Δ with the eigenvalue λ_k. Put $\dim V_k = m(k)+1$. For each integer $k\geqq 1$, let $\{ f_0, f_1, \dots, f_{m(k)} \}$ be an orthonormal basis of V_k with respect to the inner product $(f,g) = \int_M f(x)g(x)d\mu$

with the canonical measure $d\mu$ induced from g_0. Then the mapping ϕ_k of M into $\mathbb{R}^{m(k)+1}$ defined by

$$\phi_k : M \ni x \longmapsto C\cdot(\, f_0(x),\, f_1(x),\, \dots ,\, f_{m(k)}(x)\,) \in \mathbb{R}^{m(k)+1},$$

where $C = (\,Vol(M,g_0)/m(k)+1\,)^{1/2}$, gives a full G-equivariant minimal isometric immersion of $(M,(\lambda_k/n)g_0)$ into the unit sphere $S^{m(k)}(1)$(cf. Takahashi [11]). Here we call that an isometric immersion $\phi : (M,g) \to S^m(1)$ is <u>full</u> if the image $\phi(M)$ is not contained in any great hypersphere of $S^m(1)$. We call ϕ_k <u>the k-th standard minimal immersion</u> of M.

The following lemma is fundamental for the rigidity problem of the standard minimal immersion.

<u>Lemma 1</u>.(Wallach [16]) Let ψ be a minimal isometric immersion of a compact irreducible symmetric space (M,g) into $S^\ell(1)$, where g is a G-invariant Riemannian metric on $M = G/K$. If ψ is full, then
(1) there is an integer $k \geqq 1$ such that $g = c\cdot g_0$ where $c = \lambda_k/n$,
(2) $\ell \leqq m(k)$, and
(3) there is a linear mapping $A : \mathbb{R}^{m(k)+1} \longrightarrow \mathbb{R}^{m(k)+1}$ such that $\nu\cdot\psi$ is equivalent to $A\cdot\phi_k$ where ν is the natural inclusion of $S^\ell(1)$ into $S^{m(k)}(1)$.

We denote by α the second fundamental form of ψ and by σ its square length. In the research of minimal submanifolds in spheres, σ is an important intrinsic invariant.

<u>Lemma 2</u>. Under the assumption of Lemma 1, σ is given by
$\sigma = n(n-1) - (n^2/2\lambda_k)$.

Proof. It is not difficult to show that the scalar curvature of (M,g_0) is $n/2$. Hence the scalar curvature of $(M,(\lambda_k/n)g_0)$ is $n^2/2\lambda_k$. By Gauss equation and the minimality of ψ we get Lemma 2.

q.e.d.

Remark. Since λ_k is a rational number, σ is also so.

<u>Definition</u>. Let ψ be a full isometric immersion of (M,g) into $S^m(1)$. Then ψ is said to be <u>linearly rigid</u> if whenever $A : \mathbb{R}^{m+1} \longrightarrow \mathbb{R}^{m+1}$ is a linear mapping such that
(1) $A(\psi(M)) \subset S^m(1)$, and
(2) $A\cdot\psi : (M,g) \longrightarrow \mathbb{R}^{m+1}$ is an isometric immersion,

then A is orthogonal.

Proposition 3. Let $M = G/K$ be a compact irreducible symmetric space and ϕ_k be the k-th standard minimal immersion of M. Then ϕ_k is rigid if and only if ϕ_k is linearly rigid.

Now we explane the notion of the degree of an isomtric immersion $\phi : (M,g) \longrightarrow S^m(1)$. Let α be the second fundamental form of ϕ. Put $R_1 = M$ and $O_x^1 M = T_x M$. For $x \in M$, we put $O_x^2 M = \mathrm{span}\{X, \alpha(X_1,X_2) ; X, X_1, X_2 \in T_x M\}$. Let $R_2 = \{ x \in M ; \dim O_x^2 M$ is maximal $\}$. For $x \in R_2$ we put $O_x^3 M = \mathrm{span}\{ X, \alpha(X_1,X_2), (\nabla\alpha)(X_1,X_2,X_3) ; X, X_1, X_2, X_3 \in T_x M \}$. We can define $O_x^j M$, R_j $(j=2,3,\ldots)$ by recursion. Since $\dim O_x^j M \leqq m$, there is an integer $d \geqq 1$ such that $O_x^{d-1} M \subsetneqq O_x^d M = O_x^{d+1} M = \ldots$ and $R_{d-1} \supset R_d = R_{d+1} = \ldots$. Put $M' = R_d$. We call d the degree of ϕ. If $\phi(M')$ is full, then we have $O_x^d M = T_x(S^m(1))$ for any $x \in M'$ (cf. [16]). In particular $d = 1$ means that ϕ is totally geodesic. The degree of ϕ is 2 if ϕ has the parallel second fundamental form. This converse does not hold generally. There are many submanifolds with degree 2 whose second fundamental form is not parallel (cf.[9]). But the following is proved easily.

Lemma 4. Let $M = G/K$ be a symmetric space with the symmetric pair (G,K) and $\phi : G/K \longrightarrow S^m(1)$ be a G-equivariant immersion. Then the degree of ϕ is equal to 2 if and only if ϕ has the parallel second fundamental form.

The degree is related to the linearly rigidity as follows. This is a important theorem for us.

Theorem 5. (Wallach [16]) Let (M,g) be a real analytic Riemannian manifold and $\phi : (M,g) \longrightarrow S^m(1)$ be a full real analytic isometric immersion. If ϕ is of degree $\leqq 3$, then ϕ is linearly rigid.

2. Symmetric R-spaces and the canonical imbeddings

The symmetric R-space is a class of compact symmetric spaces. Each symmetric R-space has the canonical imbedding into a sphere. In this section we clarified the relation between the canonical imbedding and the standard minimal immersion in the sense of Section 1.

Let L be a noncompact connected semisimple Lie group with finite center and G be a maximal compact subgroup of L. Let $\mathfrak{l}$ and $\mathfrak{g}$ be the Lie algebras of L and G, respectively. Let $\mathfrak{l} = \mathfrak{g} + \mathfrak{p}$ be the Cartan decomposition of $\mathfrak{l}$. We denote by $B_{\mathfrak{l}}$ the Killing form of $\mathfrak{l}$. Put $\langle\ ,\ \rangle = B_{\mathfrak{l}}(\ ,\ )$. We consider $(\mathfrak{p}, \langle\ ,\ \rangle)$ as a Euclidean space. Let S be the unit sphere of $(\mathfrak{p}, \langle\ ,\ \rangle)$. We fix $v \in S$. Put $K = \{ a \in G \ ; \ Ad(a)v = v \}$. We have a compact homogeneous space $M = G/K$ and its imbedding $\Phi : G/K \longrightarrow S$ by $ao \longmapsto Ad(a)v$ where $o = eK \in G/K$. We call $M = G/K$ an R-space and Φ its canonical imbedding. If (G,K) becomes a symmetric pair, then we call $M = G/K$ a symmetric R-space. Symmetric R-spaces are completely classified (cf. Nagano [6], Kobayashi and Nagano [4]). The following is the complete table of symmetric R-spaces with L simple (cf. Takeuchi [12]) :

L	$M = G/K$
$SL(p+q;\mathbb{R})$	$G_{p,q}(\mathbb{R}) = SO(p+q)/S(O(p)\times O(q))$
$SU^*(2p+2q)$	$G_{p,q}(\mathbb{H}) = Sp(p+q)/Sp(p)\times Sp(q)$
$SU(n,n)$	$U(n)$
$SO_0(n,n)$	$SO(n)$
$SO_0(p+1,q+1)$	$Q_{p,q}(\mathbb{R}) = SO(p+1)\times SO(q+1)/S'(O(p)\times O(q))$
$SO^*(4n)$	$U(2n)/Sp(n)$
$Sp(n,\mathbb{R})$	$U(n)/O(n)$
$Sp(n,n)$	$Sp(n)$
$SL(p+q;\mathbb{C})$	$G_{p,q}(\mathbb{C}) = SU(p+q)/S(U(p)\times U(q))$
$SO(n+2;\mathbb{C})$	$Q_n(\mathbb{C}) = SO(n+2)/SO(n)\times SO(2)$
$Sp(n;\mathbb{C})$	$Sp(n)/U(n)$
$SO(2n;\mathbb{C})$	$SO(2n)/U(n)$
E_6^1	$G_{2,2}(\mathbb{H})/\mathbb{Z}_2 = Sp(4)/Sp(2)\times Sp(2)\cdot\mathbb{Z}_2$
E_6^4	$P_2(\mathbb{C}ay) = F_4/Spin(9)$
E_7^1	$SU(8)/Sp(4)\cdot\mathbb{Z}_2$
E_7^3	$T^1\cdot E_6/F_4$
$E_6^{\mathbb{C}}$	$E_6/Spin(10)\cdot T^1$
$E_7^{\mathbb{C}}$	$E_7/E_6\cdot T^1$

Remark. (1) $S'(O(p)\times O(q))$ is the subgroup of $SO(p+1)\times SO(q+1)$ consisting of matrices of the form $\begin{pmatrix} \varepsilon & 0 & & \\ 0 & A & & \\ & & \varepsilon & 0 \\ & & 0 & B \end{pmatrix}$, $\varepsilon = \pm 1$, $A \in O(p)$, $B \in O(q)$.

(2) $Sp(2)\times Sp(2)\cdot \mathbb{Z}_2$ is the subgroup of $Sp(4)$ generated by $Sp(2)\times Sp(2)$ and $\begin{pmatrix} 0 & 1_4 \\ -1_4 & 0 \end{pmatrix}$.

(3) $Sp(4)\cdot \mathbb{Z}_2$ is the subgroup of $SU(8)$ generated by $Sp(4)$ and $\begin{pmatrix} 1_4 & 0 \\ 0 & -1_4 \end{pmatrix}$.

Some properties for the canonical imbeddings are known as follows.

Proposition 6.

(1) (Takeuchi and Kobayashi [13]) If L is simple and $M = G/K$ is a symmetric R-space, then $\Phi(M)$ is a minimal submanifold in S.

(2) (Ferus [2]) If $M = G/K$ is a symmetric R-space, then Φ has the parallel second fundamental form. In the case the degree of Φ is 2.

Remark.(1) In general each canonical imbedded R-space $\Phi(M)$ is a submanifold with parallel mean curvature vector and the degree of Φ is 2 (cf.[3],[9]).

(2) The square length σ of the second fundamental form of each symmetric R-space imbedded canonically was computed by Nagura [7].

The following is our main result in this section.

Theorem 7. If $M = G/K$ is a symmetric R-space with the irreducible isotropy representation, then the canonical imbedding of M as a symmetric R-space is equivalent to the first standard minimal immersion of M in the sense of Section 1.

In the case of (1) in Proposition 6 furthermore they showed that for many symmetric R-spaces the canonical imbedding is a minimal isometric immersion by the first eigenfunctions of the Laplace-Beltrami operator with repect to the induced Riemannian metric. And they conjectured that this is true in general for all symmetric R-spaces. In this section we will give the answer to this conjecture.

Lemma 8. Let $M = G/K$ be a symmetric R-space with L simple and g_1 be the G-invariant Riemannian metric on M induced from $\langle\ \ \rangle$ through the canonical immbedding Φ. We denote by λ'_k the k-th eigenvalue of the Laplace - Beltrami operator of (M,g_1) and by $m'(k)$ the multiplicity of λ_k'. Then,

(1) $M = Q_{p,q}(\mathbb{R})$ $(p \leqslant q)$ $(\dim M = p + q, \dim \mathfrak{p} = (p + 1)(q + 1))$:

In case of $p = q$ or $q - 1$, $\lambda_1' = p + q$, $m'(1) = (p + 1)(q + 1)$.

In case of $p = q - 2$, $\lambda_1' = p + q$,

$m'(1) = (p + 1)(q + 1) + p(p + 3)/2$.

In case of $p < q - 2$, $\lambda_1' = 2(p + 1)$, $m'(1) = p(p + 3)/2$.

$\lambda_2' = p + q$, $m'(2) = (p + 1)(q + 1)$.

(2) $M = U(2n)/Sp(n)$ $(\dim M = n(2n - 1), \dim \mathfrak{p} = 2n(2n - 1))$:

$\lambda_1' = n^2$, $m'(1) = 2$.

$\lambda_2' = n(2n - 1)$, $m'(2) = 2n(2n - 1)$.

(3) $M = T^1 \cdot E_6/F_4$ $(\dim M = 27, \dim \mathfrak{p} = 54)$:

$\lambda_1' = 9$, $m'(1) = 2$.

$\lambda_2' = 27$, $m'(2) = 54$.

(4) $M = U(n)$ $(\dim M = n^2, \dim \mathfrak{p} = 2n^2)$:

$\lambda_1' = n^2$, $m'(1) = 2 + 2n^2$.

(5) $M = U(n)/O(n)$ $(\dim M = n(n + 1)/2, \dim \mathfrak{p} = n(n + 1))$:

$\lambda_1' = n(n + 1)/2$, $m'(1) = n(n + 1)$.

(6) M is ortherwise :

$\lambda_1' = \dim M$, $m'(1) = \dim \mathfrak{p}$.

First using this Lemma we prove Theorem 7.

Proof of Theorem 7. We note that in (6) of Lemma 7 each M has the irreducible isotropy representation. Let ϕ_1 be the first standard minimal immersion of M into $S^{m(1)}(1)$. ϕ_1 is a minimal isometric immersion of $(M,(\lambda_1/n)g_0)$ into $S^{m(1)}(1)$. By Lemma 8 we have $m(1) + 1 = \dim \mathfrak{p}$. Hence the canonical imbedding Φ is a minimal isometric imbedding of (M,g_1) into $S^{m(1)}(1)$. By (1) of Lemma 3, we have $g_1 = (\lambda_k/n)g_0$ for some integer $k \geqslant 1$. By the well-known theorem [11] and (6) of Lemma 8 each component function of Φ is the first eigenfunction of (M,g_1). From this we have $g_1 = (\lambda_1/n)g_0$. Φ and ϕ_1 are two minimal isometric immersions of $(M,(\lambda_1/n)g_0)$ into $S^{m(1)}(1)$. Since L is simple, Φ is full. Hence the component functions of Φ constitute a basis of the first eigenspace of M. Thus there is a linear mapping $A : \mathfrak{p} \longrightarrow \mathfrak{p}$ such that $\phi_1 = A \cdot \Phi$. Since the degree of Φ is 2, by Theorem 5 A is orthogonal. q.e.d.

We will mention the outline of the proof of Lemma 8. First we review some results from the theory of spherical functions on compact symmetric spaces (cf. Takeuchi [14]).

Let $M = G/K$ be a compact symmetric space with the symmetric pair

(G,K). Here G is a compact connected Lie group and we do not assume that G is semisimple. Let $\mathfrak{g} = \mathfrak{k} + \mathfrak{m}$ be the canonical decomposition of the Lie algebra $\mathfrak{g}$ of G and $\mathfrak{a}$ be a maximal abelian subspace of $\mathfrak{m}$. We fix an AdG-invariant inner product $(\, , \,)$ of $\mathfrak{g}$. Let $\mathfrak{t}$ be a maximal abelian subalgebra of $\mathfrak{g}$ containing $\mathfrak{a}$ and then we have $\mathfrak{t} = \mathfrak{h} + \mathfrak{a}$ where $\mathfrak{h} = \mathfrak{t} \cap \mathfrak{k}$. We fix a σ-linear order $<$ on $\mathfrak{t}$. We denote by T the maximal torus of G generated by $\mathfrak{t}$. For $\alpha \in \mathfrak{t}$, we put

$$\tilde{\mathfrak{g}}_\alpha = \{X \in \mathfrak{g}^{\mathbb{C}} \,;\, (\mathrm{ad}H)X = 2\pi\sqrt{-1}(\alpha,H)X \text{ for any } H \in \mathfrak{t}\}.$$

A element α in $\mathfrak{t}$ is called a root of $\mathfrak{g}$ with respect to $\mathfrak{t}$ if $\tilde{\mathfrak{g}}_\alpha$ is nonzero. We denote by $\Sigma(G)$ (resp. $\Sigma^+(G)$) the set of all roots (resp. positive roots) of $\mathfrak{g}$ with respect to $\mathfrak{t}$. We have the root decomposition $\mathfrak{g}^{\mathbb{C}} = \mathfrak{t}^{\mathbb{C}} + \Sigma_{\alpha\in\Sigma(G)}\tilde{\mathfrak{g}}_\alpha$. Put

$$\Gamma(G) = \{H \in \mathfrak{t} \,;\, \exp H = e\},$$
$$Z(G) = \{\lambda \in \mathfrak{t} \,;\, (\lambda,H) \in \mathbb{Z} \text{ for any } H \in \Gamma(G)\}$$
and
$$D(G) = \{\lambda \in Z(G) \,;\, (\lambda,\alpha) \geqslant 0 \text{ for any } \alpha \in \Sigma^+(G)\}.$$

Let $\mathfrak{D}(G)$ be the complete set of inequivalent irreducible unitary representations of G. Then for any $(V,\rho) \in \mathfrak{D}(G)$ the highest weight λ_ρ of (V,ρ) belongs to $D(G)$ and the mapping $\mathfrak{D}(G) \ni (V,\rho) \longmapsto \lambda_\rho \in D(G)$ is bijective.

Let A be the torus of G generated by $\mathfrak{a}$ and $\hat{A} = Ao$ be a maximal torus of $M = G/K$ where o is the origin $\{K\}$ of $M = G/K$. For $\gamma \in \mathfrak{a}$, we put

$$\mathfrak{g}^{\mathbb{C}}_\gamma = \{X \in \mathfrak{g}^{\mathbb{C}} \,;\, (\mathrm{ad}H)X = 2\pi\sqrt{-1}(\gamma,H)X \text{ for any } H \in \mathfrak{a}\}.$$

An element $\gamma \in \mathfrak{a}$ is called a root of $\mathfrak{g}$ with respect to $\mathfrak{a}$ if $\mathfrak{g}^{\mathbb{C}}_\gamma$ is nonzero. We denote by $\Sigma(G,K)$ (resp. $\Sigma^+(G,K)$) the set of all roots (resp. positive roots) of $\mathfrak{g}$ with respect to $\mathfrak{a}$. We have the decomposition $\mathfrak{g}^{\mathbb{C}} = \mathfrak{g}^{\mathbb{C}}_0 + \Sigma_{\gamma\in\Sigma(G,K)}\mathfrak{g}^{\mathbb{C}}_\gamma$. Put

$$\Gamma(G,K) = \{H \in \mathfrak{a} \,;\, \exp H\cdot o = o\},$$
$$Z(G,K) = \{\lambda \in \mathfrak{a} \,;\, (\lambda,H) \in \mathbb{Z} \text{ for any } H \in \Gamma(G,K)\}$$
and
$$D(G,K) = \{\lambda \in Z(G,K) \,;\, (\lambda,\gamma) \geqslant 0 \text{ for any } \gamma \in \Sigma^+(G,K)\}.$$

Then we have $Z(G,K) \subset Z(G)$ and $D(G,K) \subset D(G)$.

Let $\mathfrak{D}(G,K)$ be the complete set of inequivalent unitary class one representations of pair (G,K). Then for $(V_\rho,\rho) \in \mathfrak{D}(G,K)$ $(V_\rho)_K = \{v \in V_\rho : \rho(k)v = v \text{ for any } k \in K\}$ is of complex dimension 1 and the bijection $\mathfrak{D}(G) \longrightarrow D(G)$ induces the bijection $\mathfrak{D}(G,K) \longrightarrow D(G,K)$.

Proposition 9. Let g be the G-invariant Riemannian metric on M induced by $(\, , \,)$ and Δ be the Laplace-Beltrami operator of (M,g). Then

(1) the complete set of eigenvalues of Δ is given by

$$\{-a_\rho = 4\pi^2(\lambda_\rho + 2\delta(G),\lambda_\rho) \,;\, \rho \in \mathfrak{D}(G,K)\}.$$

(2) the multiplicity of the k-th eigenvalue μ_k of Δ is given by $\Sigma_\rho d_\rho$ where the summation runs over all $\rho \in \mathcal{D}(G,K)$ such that $\mu_k = -a_\rho$, and $d_\rho = \dim (V_\rho,\rho) = \Pi_{\alpha\in\Sigma^+(G)}(\alpha,\lambda_\rho + \delta(G))/(\alpha,\delta(G))$.

Outline of the proof of Lemma 8. We devide the proof into three cases :

Case 1. L is of classical type,

Case 2. L is of exceptional type and M has the irreducible isotropy representation,

Case 3. L is of exceptional type and M has the reducible isotropy representation.

Case 1 : By direct computations we can determine $\Gamma(G,K)$, $Z(G,K)$ and $D(G,K)$, individually. If M is simply connected, then a generator of $D(G,K)$ can be computed from Satake diagram of (G,K). We omit the detail of computations. By means of Proposition 9 it is not difficult to compute the first or second eigenvalue and its multiplicity.

Case 2 : (L,M) is one of the following spaces : $(E_6^1, G_{2,2}(\mathbb{H})/\mathbb{Z}_2)$, $(E_6^4, P_2(\mathbb{C}ay))$, $(E_7^1, SU(8)/Sp(4)\cdot\mathbb{Z}_2)$, $(E_6^{\mathbb{C}}, E_6/Spin(10)\cdot T^1)$ and $(E_7^{\mathbb{C}}, E_7/E_6\cdot T^1)$. By means of irreducibility of the isotropy representation, a G-invariant Riemannian metric on M is unique up to a constant factor. From the square length of the second fundamental form in the table of [7,p.222] we can find the Riemannian metric on M induced by the canonical imbedding Φ. Then by the similar method to Case 1 we can determine the first eigenvalue and its multiplicity, individually.

Case 3 : $(L,M) = (E_7^3, T^1\cdot E_6/F_4)$. Using the results of Yokota [17] we can make a concrete description of the canonical imbedding Φ of M. We consider an isometric covering $\nu : S^1 \times E_6/F_4 \longrightarrow T^1\cdot E_6/F_4$ of order 3. First we compute eigenvalues of E_6/F_4. Each eigenvalue of $S^1 \times E_6/F_4$ is the sum of an eigenvalue of S^1 and an eigenvalue of E_6/F_4. It is not difficult to determine the the first and second eigenvalues and their multiplicities of $T^1\cdot E_6/F_4$ through $\nu : S^1 \times E_6/F_4 \longrightarrow T^1\cdot E_6/F_4$.

3. Main results

Next we consider the following four compact irreducible symmetric spaces : $SU(n)/SO(n)$, $SU(n)$, $SU(2n)/Sp(n)$ and E_6/F_4. Any of them is not a symmetric R-space. We denote by M one of them. But each M can be imbedded naturally into a symmetric R-space $\tilde{M}$ in the follwing way :

M		$\tilde{M}$
$SU(n)/SO(n)$	$\longrightarrow$	$U(n)/O(n)$
$SU(n)$	$\longrightarrow$	$U(n)$
$SU(2n)/Sp(n)$	$\longrightarrow$	$U(2n)/Sp(n)$
E_6/F_4	$\longrightarrow$	$T^1 \cdot E_6/F_4$.

We denote by ι this imbedding and let Φ be the canonical imbedding of the symmetric R-space $\tilde{M}$. We consider the imbedding $\phi = \Phi \cdot \iota$ of M into the unit sphere S. For a concrete description of ϕ we refer to Naitoh [8]. By simple computations we can show the following.

Lemma 10.
(1) ϕ is a minimal immersion.
(2) The degree of ϕ is equal to 3.
(3) The square length σ of the second fundamental form of ϕ is given by

M	σ
$SU(n)/SO(n)$	$(n-1)(n+2)(n-2)(n+4)/8$
$SU(n)$	$(n-1)(n+1)(n-2)(n+2)/2$
$SU(2n)/Sp(n)$	$(n-1)(2n+1)(n+1)(n-2)$
E_6/F_4	182 .

By the same method as Theorem 7, we can verify the following.

Lemma 11. ϕ is the first standard minimal immersion of M.

Last we consider the real Grassmannian manifold of oriented p-planes in $\mathbb{R}^{p+q}$: $SO(p+q)/SO(p)\times SO(q)$ where $p\geqq q\geqq 3$. Any of them is not a symmetric R-space. We put $\tilde{G}_{p,q} = SO(p+q)/SO(p)\times SO(q)$. We denote by Ψ the natural imbedding of $\tilde{G}_{p,q}$ into the unit sphere S of $\Lambda^p\mathbb{R}^{p+q}$ with the canonical inner product. By straightforward computations we can verify the following.

Lemma 12.
(1) Ψ is a minimal immersion.
(2) The degree of Ψ is equal to q .
(3) The square length σ of the second fundamental form of Ψ is $p(p-1)q(q-1)$.

On the other hand we consider the composition $\Phi \cdot \nu$ of the natural 2-fold covering $\nu : SO(p+q)/SO(p)\times SO(q) \longrightarrow SO(p+q)/S(O(p)\times O(q))$ and the canonical imbedding Φ of $SO(p+q)/S(O(p)\times O(q))$ as a symmetric

R-space. Then $\Phi \cdot \nu$ is a minimal immersion of $\tilde{G}_{p,q}$ into a unit sphere. The first standard minimal immersion of each $\tilde{G}_{p,q}$ is given as follows.

<u>Lemma 13.</u> Let ϕ_1 (resp. ϕ_2) be the first (resp. second) standard minimal immersion of $SO(p+q)/SO(p)\times SO(q)$ $(p\geqq q\geqq 3)$.

(1) In case of $(p,q) \notin \{(3,3),(4,3),(5,3),(6,3),(4,4)\}$, ϕ_1 is equivalent to $\Phi \cdot \nu$ and $m(1)+1 = (p+q-1)(p+q+2)/2$. ϕ_2 is equivalent to Ψ and $m(2)+1 = \binom{p+q}{q}$

(2) In case of $(p,q) = (3,3)$ (resp. $(4,3),(5,3)$), ϕ_1 is equivalent to Ψ and $m(1)+1 = 20$ (resp. 35,56). ϕ_2 is equivalent to $\Phi\cdot\nu$ and $m(2)+1 = 20$ (resp. 27,35).

(3) In case of $(p,q) = (6,3)$ (resp. $(4,4)$), ϕ_1 is equivalent to $(\cos\theta_0\cdot\Phi\cdot\nu,\ \sin\theta_0\cdot\Psi)$, where $\cos\theta_0 = (\ell_1+1/\ell_1+\ell_2+2)^{1/2}$ and $\sin\theta_0 = (\ell_2+1/\ell_1+\ell_2+1)^{1/2}$, and $m(1)+1 = 128$ (resp. 105). ℓ_1 and ℓ_2 are given by

	(6,3)	(4,4)
ℓ_1+1	44	35
ℓ_2+1	84	70

.

In this case $\Psi_\theta = (\cos\theta\cdot\Psi\cdot\nu,\ \sin\theta\cdot\Psi)$ $(0\leqq\theta\leqq\pi/2)$ constitutes a family of mutually inequivalent minimal isometric immersion of $(SO(p+q)/SO(p)\times SO(q),(\lambda_1/pq)g_0)$ into $S^{m(1)}(1)$.

Proof of Theorem A. By means of Theorem 7, Proposition 6(2), Lemma 11, Lemma 10(3), Lemma 13(1)(2) and Theorem 5, ϕ_1 is linearly rigid. By Proposion 3 ϕ_1 is rigid. q.e.d.

Theorem B is a direct result from (3) of Lemma 13.

Remark. (1) Though the first standard minimal immersion of $SO(9)/SO(6)\times SO(3)$ (resp. $SO(8)/SO(4)\times SO(4)$) is nonrigid, the first standard minimal immersion of $SO(9)/S(O(6)\times O(3))$ (resp. $SO(8)/S(O(4)\times O(4))$) is rigid.

(2) In other words Case (3) show that the first eigenspace of $SO(9)/SO(6)\times SO(3)$ (resp. $SO(8)/SO(4)\times SO(4)$) in the space of all real-valued C^∞-functions decomposes into two irreducible $SO(9)$- (resp. three irreducible $SO(8)$-) modules over $\mathbb{R}$.

4. <u>Remarks and problems</u>

Further it is interesting to study the rigidity of standard minimal immersions with higher order. We can know the rigidity of the second standard minimal immersion for a few spaces.

Proposition 14. Let M be one of the following : $Sp(4)/Sp(2)\times Sp(2)$, $SU(8)/Sp(4)$, and $SO(p+3)/SO(p)\times SO(3)$ ($p\geqq 3, p\neq 6$). Then the second standard minimal immersion of M is rigid.

On the other hand by the similar argument to the proof of Theorem A we have the following.

Proposition 15. Let M be one of the following symmetric R-spaces : $Q_{p,q}(\mathbb{R})$ ($p=q, q-1$ or $<q-2$), $U(2n)/Sp(n)$, $T^1\cdot E_6/F_4$ and $U(n)/O(n)$. Then the canonical imbedding Φ of M is rigid as a minimal isometric immersion.

Finally we ask the following :

Problem. For any other compact irreducible symmetric space M of exceptional type, is the first standard minimal immersion of M rigid ?

Added in proof. After the author finished this work, he knew that first eigenvalues of symmetric R-spaces were determined in Takeuchi's preprint [15]. The simple and unified proof of Lemma 8 is given in the paper.

References

[1] M.do Carmo and N.Wallach, Minimal immersions of spheres into spheres , Ann. of Math. 95 (1971), 43-62.

[2] D.Ferus, Immersionen mit paralleler zweiter Fundamentalform : Beispiele und Nicht-Beispiele, Manuscripta Math. 12 (1974), 153-162.

[3] Y.Kitagawa and Y.Ohnita, On the mean curvature of R-spaces, Math. Ann. 262 (1983), 239-243.

[4] S.Kobayashi and T.Nagano, On filtered Lie algebras and geometric structures I, J.Math.Mech. 13 no.5 (1964), 875-907.

[5] K.Mashimo, Degree of the standard isometric minimal immersions of the symmetric spaces of rank one into spheres, Tsukuba J.Math. 5 (1981), 291-297.

[6] T.Nagano, Transformation groups on compact symmetric spaces, Trans. Amer.Math.Soc. 118 (1965), 428-453.

[7] T.Nagura, On the lengths of the second fundamental forms of R-spaces

, Osaka J.Math. 14 (1977), 207-223.

[8] H.Naitoh, Totally real parallel submanifolds in $P^n(c)$, Tokyo J.Math. 4 no.2 (1981), 291-306.

[9] Y.Ohnita, The degrees of the standard imbeddings of R-spaces, Tohoku Math.J. 35 (1983), 499-502.

[10] K.Sakamoto, Planar geodesic immersions, Tohoku Math.J. 29 (1977), 25-56.

[11] T.Takahashi, Minimal immersions of Riemannian manifolds, J.Math.Soc. Japan 18 (1966), 380-385.

[12] M.Takeuchi, Cell decompositions and Morse equalities on certain symmetric spaces, J.Fac.Sci.Univ.Tokyo, Ser I, 12 (1965) 81-192.

[13] M.Takeuchi and S.Kobayashi, Minimal imbeddings of R-spaces, J.Diff.Geom. 2 (1968), 203-215.

[14] M.Takeuchi, Modern theory of spherical functions (in Japanese), Iwanami, Tokyo, 1975.

[15] M.Takeuchi, On stability of certain minimal submanifolds of compact Hermitian symmetric spaces, preprint.

[16] N.Wallach, Minimal immersions of symmetric spaces into spheres, in Symmetric Spaces (W.B.Boothby and G.L.Weiss,eds.), Marcel Dekker, New Nork, 1972, 1-40.

[17] I.Yokota, Simply connected compact simple Lie groups $E_{6(-78)}$ of type E_6 and its involutive automorphisms, J.Math.Kyoto Univ. 20-3 (1980), 447-473.

A VARIATIONAL APPROACH TO THE BOUNDARY VALUE PROBLEM FOR HYPERSURFACES WITH PRESCRIBED MEAN CURVATURE

Shôichirô TAKAKUWA

Department of Mathematics
Tokyo Institute of Technology
Oh-okayama, Meguro-ku, Tokyo, 152,
Japan

§1. Introduction.

In this article we discuss the problem to find a surface having a prescribed mean curvature and a given boundary value. We shall approach to this problem by the variational method. Let U be an open set in R^n and A(x) be a function on U. For a fixed hypersurface Γ in R^n we define the family of hypersurfaces

$$\Xi = \{ S \text{ is a hypersurface in } R^n \; ; \; S = \Gamma \text{ outside } U \} ,$$

and define the functional on Ξ as follows.

$$I_A(S) = \mathrm{Area}(S \cap U) + \int_{V(S)\cap U} A(x)dx ,$$

where V(S) denotes a subset bounded by S. It is known that an extremal for the functional I_A on Ξ is a hypersurface whose mean curvature is prescribed by A(x) and boundary value is given by Γ.

The traditional approach to variational problem is to consider a surface as a mapping from a suitable manifold M into R^n. However, this approach encounters a difficulty since the functional I_A is invariant under the action of the large group, that is, the group of diffeomorphisms of M. Furthermore, many other difficulties are coming from considering sufaces as mappings (see [1]). For these reasons the traditional approach has failed to give significant results for functionals as I_A.

While, the measure-theoretic approaches have been studied by several authors, e.g. de Giorgi, Federer, Almgren since 1950's (see [1], [2], [7]). They have overcomed the above difficulties and yielded considerably satisfactory results. In this article we shall state the results for the functional I_A obtained by one of the measure-theoretic methods, in which functions of bounded variation play the crucial role.

Our program is as follows. Section 2 is devoted to give a brief survey of the space of functions of bounded variation. In Section 3 we give an outline of the variational approach to the problem and state the existence and regularity results. In Section 4 we concentrate our interest on a special case called the non-parametric problem. In the final section we present some remarks of the non-parametric problem, which are concerned with the results different from general ones described in Section 3.

§2. Space of functions of bounded variation.

Let U be an open subset of R^n ($n \geq 2$). We define by BV(U) the space of all funcions of bounded variation in U, that is, $u \in BV(U)$ if and only if $u \in L^1(U)$ and its gradient $\nabla u = (\partial u/\partial x_1, \cdots\cdots, \partial u/\partial x_n)$, in the sense of distributions, is a vector-valued Radon measure with a finite total variation on U. BV(U) is a Banach space under the norm

$$(2.1) \qquad ||u||_{BV(U)} = ||u||_{L^1(U)} + \int_U |\nabla u| ,$$

where $\int_U |\nabla u|$ denotes the total variation of the vector-valued measure ∇u on U. And in case that the boundary ∂U is a Lipschitz manifold, the trace $\gamma(u)$ ($u \in BV(U)$) on the boundary ∂U is defined and belongs to $L^1(\partial U)$ (see §2 in [7]).

The Sobolev space $W^{1,1}(U)$ is contained in BV(U) and for each $u \in W^{1,1}(U)$ we have

$$(2.2) \qquad \int_U |\nabla u| = \int_U |\nabla u(x)|dx \quad \text{and} \quad ||u||_{BV(U)} = ||u||_{W^{1,1}(U)}.$$

This formula indicates that $W^{1,1}(U)$ is a closed subspace of BV(U).

Another example of functions of BV(U) is described as follows.

<u>2.1. Example.</u> Suppose that an open set E in R^n has C^2 boundary. We consider χ_E, the characteristic function of E, which is defined by

$$\chi_E(x) = 1 \text{ if } x \in E, \qquad \chi_E(x) = 0 \text{ if } x \in R^n - E.$$

Then, for any bounded open set U in R^n we have

$$(2.3) \qquad \chi_E \in BV(U) \quad \text{and} \quad \int_U |\nabla \chi_E| = H_{n-1}(U \cap \partial E) ,$$

where H_{n-1} denotes the (n-1)-dimensional Hausdorff measure. However, χ_E does not belong to the Sobolev space $W^{1,1}(U)$.

The above results are extended to a more general class of sets in R^n.

<u>2.2. Definition.</u> (§3 in [7]) Let E be a Borel subset and U be a subset of R^n.

(1) Define the <u>perimeter of E in U</u> as

$$(2.4) \qquad P(E,U) = \int_U |\nabla \chi_E| .$$

In case $U = R^n$ we abbreviate $P(E) = P(E,R^n)$.

(2) A Borel set E is called a Caccioppoli set if and only if E has a locally finite perimeter, that is, $P(E,U) < \infty$ for each bounded open set U in R^n.

Caccioppoli sets have the following property.

2.3. Proposition. (§1. of [7]) Let E be a Caccioppoli set in R^n.

(1) For each open set U in R^n, we have

(2.5) $$P(E,U) = \int_U |\nabla \chi_E| \leq H_{n-1}(\partial E \cap U).$$

(2) The support of the vector-valued measure χ_E is contained in ∂E.

(3) (generalized Gauss-Green formula)

(2.6) $$\int_E \operatorname{div} G\, dx = - \int_{\partial E} G \cdot \nabla \chi_E \qquad \text{for any } G \in C_0^1(R^n, R^n).$$

Furthermore, it is known that the Gauss-Green formula stated above characterizes Caccioppoli sets conversely (see §1.8 of [7]).

§3. Variational problem for hypersurface with prescribed mean curvature.

In this section we present a brief survey of a variational approach to the problem to find a hypersurface having a prescribed mean curvature and boundary value. We first discuss the existence of generalized solutions for the variational problem. We intend to seek its generalized solutions as the boundary of Caccioppoli sets, because Proposition 2.3 assures that the boundary of a Caccioppoli set may be regarded as a 'generalized surface'.

We formulate the variational problem as follows. Let U be an open set in R^n, L be a fixed Caccioppoli set and A(x) be a function belonging to $L^1(U)$. We define the family of Caccioppoli sets as

$$\Sigma = \{ E \subset R^n \ ; \ E \text{ is a Caccioppoli set with } E - \overline{U} = L - \overline{U} \}.$$

On Σ we consider the functional

$$I_A(E) = P(E,\overline{U}) + \int_U A(x)\chi_E(x)dx \qquad \text{for } E \in \Sigma. \tag{3.1}$$

The lower boundedness of I_A follows immediately, that is,

$$I_A(E) \geq - \int_U |A(x)|dx > -\infty \qquad \text{for any } E \in \Sigma. \tag{3.2}$$

On the variational problem for the functional I_A on Σ, we state the following existence theorem.

3.1. Theorem. ([8], [9]) *Suppose that* U *is a bounded open set in* R^n *and* L, A *are defined as above. Then, there exists* $E_0 \in \Sigma$ *such that*

$$I_A(E_0) = \inf \{ I_A(E) \ ; \ E \in \Sigma \}. \tag{3.3}$$

This theorem is obtained by applying the theorem on lower semi-continuity of I_A with respect to L^1 convergence and the Rellich-Kondrachov theorem concerning with compactness of the Sobolev imbedding $BV(U) \longrightarrow L^1(U)$.

We next discuss the regularity of the solution E_0, which is whether the boundary ∂E_0 of E_0 is actually a sufficiently smooth hypersurface in R^n.

3.2. Theorem. ([8], [9]) *Suppose that* U, A, L *are stated as above and moreover* $A \in L^\infty(U)$. *Let* E_0 *be the solution obtained in Theorem 3.1. Then, there exists an open subset* W *of* U *such that* $\partial E_0 \cap W$ *is a hypersurface of class* $C^{1,\alpha}$ (*for some* α, $0 < \alpha < 1$) *and*

(3.4) $$H_s(U - W) = 0 \quad \underline{\text{for any}}\ s > n - 8$$

holds.

§4. Non-parametric problem.

In this section we discuss a special problem to seek a solution over hypersurfaces represented as the graph of a function, which is often called the non-parametric problem. Let Ω be a domain in R^n with Lipschitz boundary, $U = \Omega \times R$ and $A(x,t)$ be a function on U. For each function u on Ω its graph G_u over Ω is a hypersurface in U and the value of the functional I_A for G_u is written as follows.

$$J_A(u) \equiv I_A(G_u) = \int_\Omega (1 + |\nabla u|^2)^{1/2} dx + \int_\Omega \int_0^{u(x)} A(x,t)\,dt\,dx.$$

We can easily deduce that solutions of the non-parametric problem are presented by extremals for the functional J_A over all functions on Ω taking a fixed boundary value ϕ on $\partial\Omega$. And the extremals are characterized by the following Euler-Lagrange equation.

$$\operatorname{div}\{\nabla u/(1 + |\nabla u|^2)^{1/2}\} = A(x,u) \quad \text{in } \Omega, \quad u = \phi \quad \text{on } \partial\Omega.$$

In this article we shall state the results only for a simple example

$(*)_T$ $$\operatorname{div}\{\nabla u/(1 + |\nabla u|^2)^{1/2}\} = -T\,c(x) \quad \text{in } \Omega, \quad u = \phi \quad \text{on } \partial\Omega,$$

where c is a function on Ω and T is a positive constant.

The associated variational problem is to find a function in $BV(\Omega)$ which minimizes the functional

(4.1) $$J_T(u) = \int_\Omega (1 + |\nabla u|^2)^{1/2} - T \int_\Omega c\,u\,dx + \int_{\partial\Omega} |\gamma(u) - \phi|\,dH_{n-1},$$

on BV(Ω). Here, $\int_\Omega (1 + |\nabla u|^2)^{1/2}$ is the total variation of a vector-valued measure (dx, $\partial u/\partial x_1, \cdots\cdots, \partial u/\partial x_n$) on Ω where dx denotes the n-dimensional Lebesgue measure. One of remarkablefacts for the non-parametric problem is that the lower boundedness of the functional breaks in genaral. As an example we state the following theorem.

4.1. Theorem. ([10], [12]) *Suppose that Ω is a bounded domain in R^n, c is a non-negative function in $L^\infty(\Omega)$ and $\phi \in L^1(\partial\Omega)$. Then, the necessary and sufficient condition for J_T to be bounded from below on BV(Ω) is that*

$$(4.2) \qquad T \leq T^* = \inf\{ H_{n-1}(\partial E)/\text{meas}_c(E) \; ; \; E \subset \Omega \},$$

where $\text{meas}_c(E) = \int_E c(x)dx$ *and the infimum is taken over all open subsets of Ω with C^2 boundary.*

Proof. We first prove that the condition (4.2) is necessary. We assume that $T > T^*$. By the definition of T^* there exist $\lambda > 0$ and an open set $G \subset\subset \Omega$ with C^2 boundary such that

$$T > (H_{n-1}(\partial G) + \lambda)/\text{meas}_c(G).$$

We define

$$u_j(x) = j \cdot \chi_G(x) \qquad (j = 1, 2, \cdots\cdots).$$

Then,

$$J_T(u_j) \leq j \int_\Omega |\nabla \chi_G| + \text{meas}(\Omega) - j\,T \int_\Omega c\chi_G\, dx + \|\phi\|_{L^1(\partial\Omega)},$$

$$\leq j\,(H_{n-1}(\partial G) - T\,\text{meas}_c(G)) + \text{meas}(\Omega) + \|\phi\|_{L^1(\partial\Omega)},$$

$$< - j\lambda + \text{meas}(\Omega) + \|\phi\|_{L^1(\partial\Omega)}.$$

Hence, we have

$$\mu_T = \inf_{BV(\Omega)} J_T = -\infty \qquad \text{for } T > T^*.$$

Next we prove the condition (4.2) is sufficient. We first consider $v \in C^{\infty}(\Omega)$ such that $\gamma(v) = 0$. We set

$$A(t) = \{ x \in \Omega \; ; \; |v(x)| > t \}, \qquad a_t = \chi_{A(t)} \qquad (t \geq 0).$$

By the co-area formula (§1.23 of [7]) we have

$$|v(x)| = \int_0^{\infty} a_t(x)\, dt , \qquad \int_{\Omega} |\nabla |v|| = \int_0^{\infty} \left(\int_{\Omega} |\nabla a_t| \right) dt.$$

Using Sard's theorem we observe that $\partial A(t)$ is of class C^{∞} for almost all $t > 0$. From (2.3) we obtain

$$\int_{\Omega} |\nabla |v|| = \int_0^{\infty} H_{n-1}(\partial A(t))\, dt ,$$

$$J_T(|v|) \geq \int_0^{\infty} \{ H_{n-1}(\partial A(t)) - T\, \mathrm{meas}_c(A(t)) \}\, dt \geq 0.$$

By [13] we take an extension $\tilde{\phi} \in W^{1,1}(\Omega)$ of ϕ. Then,

$$J_T(v + \tilde{\phi}) \geq J_T(|v|) - J_T(\tilde{\phi}) \geq - J_T(\tilde{\phi}).$$

For any $u \in BV(\Omega)$ with $\gamma(u) = \phi$ we apply the approximation theorem [§1.17 of 7] for $v = u - \tilde{\phi}$ to obtain

$$J_T(u) \geq - J_T(\tilde{\phi}) .$$

Using the following result [13]

$$\inf_{BV(\Omega)} J_T = \inf \{ J_T(u) \; ; \; u \in BV(\Omega), \; \gamma(u) = \phi \} ,$$

the proof is completed. q.e.d.

We next state the existence theorem of generalized solutions of this variational problem.

4.2. Theorem. ([12]) Suppose that Ω, c, ϕ are as in Theorem 4.1 and moreover $T < T^*$. Then, there exists a function $u_T \in BV(\Omega)$ such that

$$J_T(u_T) = \inf \{ J_T(u) \; ; \; u \in BV(\Omega) \} .$$

Proof. We choose a minimizing sequence $\{u_j\}$ of $BV(\Omega)$. We may assume that the sequence $\{J_T(u_j)\}$ is bounded. By virtue of Theorem 4.1 and the definition of J_T we have

$$\begin{aligned} &\int_\Omega (1 + |\nabla u_j|^2)^{1/2} + \int_{\partial\Omega} |\gamma(u_j)| dH_{n-1} \\ &\leq \frac{1}{T^* - T} (T^* J_T(u_j) - T \mu_{T^*}) + \|\phi\|_{L^1(\partial\Omega)} , \end{aligned} \tag{4.3}$$

where $\mu_T = \inf\limits_{BV(\Omega)} J_T$ for $T \leq T^*$. Next we apply the Poincaré inequality for functions of bounded variation to obtain

$$\|u_j\|_{L^1(\Omega)} \leq n(\text{meas}(\Omega)/\omega_n)^{1/n}\left(\int_\Omega |\nabla u_j| + \int_{\partial\Omega} |\gamma(u_j)| dH_{n-1}\right) \tag{4.4}$$

where ω_n denotes the volume of a unit ball in R^n. According to (4.3), (4.4) we can deduce that $\{u_j\}$ is bounded in $BV(\Omega)$.

Using the results of Ferro [3] on a weak* topology on $BV(\Omega)$ (denoted by w_q^* topology) we are able to choose a subsequence $\{u_k\}$ of $\{u_j\}$ which converges to some element u_T of $BV(\Omega)$ in the w_q^* topology. From lower semicontinuity of the functional J_T with respect to the w_q^* topology (see [3], [11]) we have

$$\mu_T \leq J_T(u_T) \leq \liminf_{k \to \infty} J_T(u_k) = \mu_T. \qquad \text{q.e.d.}$$

4.3. Remark. For $\phi \in L^1(\partial\Omega)$, there exists $\Phi \in BV(R^n)$ with $\Phi|_{\partial\Omega} = \phi$, where existence of Φ is proved in [13]. We define the set

$$E_T = \{ (x,t) \in R^{n+1} \; ; \; u_T(x) < t \;\; \text{if } x \in \Omega, \; \Phi(x) < t \;\; \text{if } x \notin \Omega \}.$$

Then, E_T is a Caccioppoli set and moreover a solution of the variational problem in Theorem 3.1 for $U = \Omega \times R$, $L = \{ (x,t) \in R^{n+1} ; \Phi(x) < t \}$ and $A(x,t) = T\,c(x)$ for $(x,t) \in R^{n+1}$.

We next discuss the regularity of the solution u_T. From the above remark and Theorem 3.2, the graph of u_T is a hypersurface of class $C^{1,\alpha}$ (for some α, $0 < \alpha < 1$) except for a set of singular point. Furthermore, the following stronger result is known.

4.4. Theorem. ([4]) Let Ω be a bounded domain in R^n with C^2 boundary, $\phi \in C^0(\partial\Omega)$ and $T < T^*$. Suppose that a non-negative function $c \in C^1(\bar{\Omega})$ satisfies

$$(4.5) \qquad T\,c(y) \leq (n-1)\,H(y) \qquad \text{for any } y \in \partial\Omega,$$

where H denotes the mean curvature of $\partial\Omega$. Then, the solution u_T obtained in Theorem 4.2 belongs to $C^{2,\alpha}(\Omega) \cap C^0(\bar{\Omega})$ (for any α, $0 \leq \alpha < 1$), $u_T = \phi$ on $\partial\Omega$ and u_T is a unique solution of Eq.$(*)_T$.

The interior regularity of u_T is derived from the removability of singular points, which is due to C^1 a priori estimate for Eq.$(*)_T$ (see [4]). The boundary regularity and the condition (4.5) are first presented in Serrin [11] by applying the maximum principle with nice choices of barrier functions.

§5. Some remarks on the non-parametric problem.

We finally state some results peculiar to the non-parametric problem. We first present the following non-existence theorem.

5.1. Theorem. (Remark 3.3 of [12]) Let Ω, c, ϕ, T^* be as in Theorem 4.1. Then, the non-parametric problem $(*)_T$ has no solution in $BV(\Omega)$ for each $T > T^*$.

Proof. We note that the functional J_T is convex and does not take any other critical values than the minimum. We combine this fact with Theorem 4.1 to obtain the required result. q.e.d.

The second remark for the non-parametric problem is that the lower boundedness of the functional does not always imply the existence of the solution. An example is stated as follows.

5.2. Theorem. (Theorem 5.3 of [12]) Let Ω be a bounded domain in R^n with C^3 boundary and $\phi \in C^{1,\beta}(\partial\Omega)$ for some $\beta > 0$. Suppose a non-negative function $c \in C^1(\bar{\Omega})$ is not identically zero and satisfies

$$T^*c(y) \leq (n - 1)\, H(y) \qquad \text{for any } y \in \partial\Omega.$$

Then, the non-parametric problem $(*)_{T^*}$ has no solution in $C^2(\Omega) \cap C^0(\bar{\Omega})$ and $\sup_\Omega u_T$ diverges as T tends to T*.

To give the proof of this theorem we state the following lemma.

5.3. Lemma. (Theorem 5.1. of [12]) Suppose that Ω, c is as above. Let $w \in C^1(\Omega)$ be a weak solution of the equation

$$- \operatorname{div}\{\nabla u/(1 + |\nabla u|^2)^{1/2}\} = T^*c \quad \text{in } \Omega.$$

Then, $\sup_\Omega |\nabla w| = \infty$, that is w does not belong to $C^1(\bar{\Omega})$.

Proof. By the definition of the weak solution and the approximation theorem [§1.17 of 7] we can easily deduce that

$$(5.1) \qquad T^*\int_\Omega c\,\eta\, dx \leq M \int_\Omega |\nabla\eta| \qquad \text{for any } \eta \in BV(\Omega) \text{ with } \gamma(\eta) = 0$$

where $M = \sup_\Omega\{|\nabla w|/(1 + |\nabla w|^2)^{1/2}\}$. We choose η as

$$\eta = \chi_E \qquad \text{for any } E \subset\subset \Omega \text{ with } C^2 \text{ boundary } \partial E.$$

Using (2.3) we have

$$T^* \leq M \cdot H_{n-1}(\partial E)/\mathrm{meas}_c(E) .$$

In the right hand of the above inequality we take infimum with respect to E so that we obtain

$$T^* \le M \cdot T^* \le T^* \ , \quad \text{that is,} \quad M = 1 \ .$$

This completes the proof. q.e.d.

Proof of Theorem 5.2. We assume that the conclusion of the theorem does not hold, that is, we assume that there exists a constant K independent of T such that

$$\sup_{\Omega} u_T \le K \qquad \text{for all } T < T^*.$$

Using the maximum principle [§9.2 of 6] we first observe that $\{u_T(x)\}$ ($T < T^*$) is monotone increasing for each $x \in \overline{\Omega}$. Hence, the limiting value

$$u_{T^*}(x) = \lim_{T \uparrow T^*} u_T(x)$$

exists for each $x \in \overline{\Omega}$ and $u_{T^*}(x) = \phi(x)$ for each $x \in \partial\Omega$.

By applying a priori estimates for Eq.$(*)_T$ (see [§12, 13, 15 of 6]) we are able to show that u_{T^*} belongs to $C^2(\Omega) \cap C^0(\overline{\Omega})$ and satisfies Eq.$(*)_{T^*}$.

Furthermore, under the regularity hypotheses for Ω, c, ϕ the result of Giaquinta [5] states that u_{T^*} is Lipschitz continuous on $\overline{\Omega}$. However, this contradicts with Lemma 5.3. Therefore, the theorem must hold. q.e.d.

For the case $T = T^*$ we have not yet obtained a condition for Ω, c, which characterizes the solvability of Eq.$(*)_{T^*}$. We have a conjecture that the above mentioned results are deeply connected with the possibility for the solution to be represented as the graph over Ω.

REFERENCES.

[1] Almgren, F. J. : "Plateau's Problem," Benjamin, New York 1966.

[2] Federer, H. : "Geometric Measure Theory," Berlin-Heidelberg-New York, Springer-Verlag 1969.

[3] Ferro, F. : Variational functional defined on spaces of BV functions and their dependence on boundary data, Ann. Mat. Pura Appl. (4) 122 (1979), 269-287.

[4] Giaquinta, M. : Regolarità delle superfici BV(Ω) con curvatura media assegnata, Boll. U. M. I. (4) 8 (1973), 567-578.

[5] Giaquinta, M. : On the Dirichlet problem for surfaces of prescribed mean curvature, Manuscripta Math. 12 (1974), 73-86.

[6] Gilbarg, D. & N. S. Trudinger, : "Elliptic Partial Differential Equations of Second Order," Berlin-Heidelberg-New York, Springer-Verlag 1977.

[7] Giusti, E. : "Minimal Surface and Functions of Bounded Variation," Notes on Pure Math. 10, Australian National Univ., Canberra 1977.

[8] Massari, U. : Esistenza e regolarità delle ipersuperfici di curvatura media assegnata in R^n, Arch. Rational Mech. Anal. 55 (1974), 357-382.

[9] Miranda, M. : Existence and regularity of hypersurfaces of R^n with prescribed mean curvature, Proc. Symp. Pure Math. 23 (1973), 1-11.

[10] Mosolov, P. P. : A functional connected with a surface of given mean curvature, Math. USSR Sbornik 7, No. 1 (1969), 45-58.

[11] Serrin, J. : The problem of Dirichlet for quasilinear elliptic differential equations with many independent variables, Philos. Trans. Roy. Soc. London Ser. A 264 (1969), 413-496.

[12] Takakuwa, S. : On a parameter dependence of solvability of the Dirichlet problem for non-parametric surfaces of prescribed mean curvature, to appear.

[13] Williams, G. H. : The equivalence of some variational problem for surfaces of prescribed mean curvature, Bull. Austral. Math. Soc. 20 (1979), 87-104.

Holomorphic embedding of compact s.p.c. manifolds into complex manifolds as real hypersurfaces

By Takeo OHSAWA

Research Institute for Mathematical Sciences,

Kyoto University, 606 Kyoto JAPAN

Let D be a relatively compact domain in a complex manifold. A sufficient condition for D to be a holomorphically convex manifold is that the boundary of D is everywhere strongly pseudoconvex (i.e., for every point $x \in \partial D$ there exist a neighbourhood $U \ni x$ and a C^2 strictly plurisubharmonic function $\varphi: U \longrightarrow \mathbb{R}$ such that $d\varphi_x \neq 0$ and $U \cap D = \{p ; \varphi(p) < 0\}$). Such a domain is called an s.p.c. domain (s.p.c. signifies strongly pseudoconvex). In [13] we obtained a characterization of s.p.c. domains in the category of weakly 1-complete manifolds (complex manifolds which admit C^∞ plurisubharmonic exhaustion function). Namely we obtained the following result.

<u>Theorem 1.</u> Let X be a complex manifold with a bounded exhaustion function φ which is C^∞ strictly plurisubharmonic outside a compact subset, and let ds^2 be a hermitian metric on X which coincides with $\partial\bar{\partial}\varphi$ outside a compact subset of X. Assume that the diameter and the volume of X are finite, that the pointwise norms of $\nabla\varphi$, $\nabla\nabla\varphi$, ..., $\nabla\nabla\cdots\nabla\varphi$, ... are bounded, and that there exists a constant $c > 0$ such that $|\partial\varphi|$, $|\partial\varphi|^{-1} < c$ outside a compact subset. If moreover $\dim_{\mathbb{C}} X \geq 3$, then X is an s.p.c. domain. Namely,

X is biholomorphic to an s.p.c. domain in some complex manifold.

In the proof of Theorem 1, the crucial step is a realization of <u>the boundary</u> of X (the complement of X in the completion X with respect to the metric ds^2) as a real hypersurface in a complex manifold. For that purpose we need the following result which will be interesting to geometers. We shall give its proof here.

<u>Theorem 2.</u> Let X be a compact s.p.c. manifold (for the definition see below) of dimension ≥ 5. Then there exist a complex manifold M and a holomorphic embedding of X into M as a real hypersurface.

§1. <u>Definitions and Known Results</u>

Let X be a (C^∞) differentiable manifold and let T_X be its tangent bundle. A CR-structure on X is a differentiable subbundle $T'_X \subset T_X \otimes \mathbb{C}$ which is closed under the operation of Poisson bracket and satisfies $T'_X \cap \overline{T}'_X = 0$. Here $\overline{T}'_X$ denotes the complex conjugate of T'_X. For a real submanifold W of a complex manifold M, $T_W \otimes \mathbb{C}$ is closed under the Poisson bracket. But it is not always the case that the rank of $(T_{W,x} \otimes \mathbb{C}) \cap T^{1,0}_{M,x}$ is constant. If it is the case, $T'_W := (T_W \otimes \mathbb{C}) \cap T^{1,0}_{M|W}$ defines a CR-structure on W. W is then called a CR-submanifold of M. X is called an s.p.c. manifold if a CR-structure T'_X on X is given which satisfies the following additional conditions:

(1) $\quad \mathrm{rank}_{\mathbb{C}} T'_X = \frac{1}{2}(\dim_{\mathbb{R}} X - 1)$

(2) $\quad$ Fix an identification $\theta \colon T_X \otimes \mathbb{C} \,/\, T'_X \oplus \overline{T}'_X \cong \mathbb{C}$ which satisfies $\theta(\overline{v}) = \overline{\theta(v)}$. Then, for any local frame $(v_1, \ldots, v_n)$ of T'_X, the matrix $(-\sqrt{-1}\,\theta([v_i, v_j]))$ is positive or

negative definite.

Note that the conditions above are satisfied by a real hypersurface S in a complex manifold if and only if S is the zero set of a C^∞ strictly plurisubharmonic function defined on a neighbourhood of S.

A $\mathbb{C}$-valued function f defined on an open set $V \subset X$ is said to be holomorphic if f is of class C^∞ and $vf = 0$ for any section v of $\overline{T}'_X$ over V. The restrictions of holomorphic functions to real hypersurfaces are holomorphic. Concerning the embedding of s.p.c. manifolds the following results are known before.

<u>Theorem B.</u> (cf.[1]) Let (X, T'_X) be a compact s.p.c. manifold of dimension $2n-1$. If $n \geqq 3$, then there exist an integer N and holomorphic functions f_i, $i=1,2,\dots,N$, such that the map $(f_1, f_2, \dots$ $\dots, f_N)$ gives an embedding of X into $\mathbb{C}^N$.

<u>Corollary 3.</u> Every compact s.p.c. manifold of dimension $\geqq 5$ is isomorphic to a CR-submanifold of a complex number space.

Proof. Let $F: X \hookrightarrow \mathbb{C}^N$ be an embedding by holomorphic functions. Then, for any point $x \in X$, $F_*(T'_{X,x}) \subset T^{1,0}_{\mathbb{C}^N, F(x)} \cap (T_{F(X)} \otimes \mathbb{C})$. Since $\operatorname{rank}_{\mathbb{C}} T'_X = n-1$, we have $F_*(T'_{X,x}) = T^{1,0}_{\mathbb{C}^N, F(x)} \cap (T_{F(X)} \otimes \mathbb{C})$.

<u>Corollary 4.</u> Every compact s.p.c. manifold of dimension $\geqq 5$ is locally holomorphically embeddable as a real hypersurface.

For our purpose Cor.3 and Cor.4 suffice, but it will be worthwhile to take notice of the following deep result due to Kuranishi.

<u>Theorem</u> (cf. [10]) Every s.p.c. manifold of dimension $\geqq 9$ is locally holomorphically embeddable as a real hypersurface.

§2. Cohomology Groups and Tanaka's Stability Theorem

We need the notion of a complex manifold with boundary. A complex manifold with boundary is a C^∞ manifold with boundary with a system of coordinate patches $b_i : U_i \xrightarrow{\sim} \{(z_1,z_2,\dots,z_n) \in \mathbb{C}^n;\ \sum_{k=1}^n |z_k|^2 < 1,\ r_i(z_1,z_2,\dots,z_n) \leqq 0\}$, where r_i is a C^∞ real valued function on the ball with $dr_i = 0$ everywhere and b_i are diffeomorphisms such that $b_j b_i^{-1}$ is holomorphic on $b_i(U_i \cap U_j) \setminus \{r_i = 0\}$.

In virtue of Theorem B, we have only to deal with s.p.c. submanifolds of $\mathbb{C}^N$. For these manifolds we have the following

Proposition 5. (cf. [11] section 2) Let $X \subset \mathbb{C}^N$ be a compact $(2n-1)$-dimensional s.p.c. submanifold. If $n \geqq 2$, then there exists a unique analytic subvariety W in $\mathbb{C}^N \setminus X$ such that the closure $\overline{W}$ in $\mathbb{C}^N$ is compact and that $\overline{W} \setminus W = X$. W has the following property: $\mathrm{Sing}W$ (the set of singular points of W) is a finite set and $\overline{W} \setminus \mathrm{Sing}W$ is a complex manifold with boundary.

Proof. Choose a finite system of open subsets $\{U_i\}_{i=1}^m$ of $\mathbb{C}^N$ and projections $p_i : \mathbb{C}^N \to \mathbb{C}^n$ so that p_i induces a holomorphic embedding of $U_i \cap X$ into $\mathbb{C}^n$. We choose $\{U_i\}$ so that $U_i \cap X$ are connected and that each $p_i(U_i \cap X)$ is contained in a ball and separates it into two connected components $p_i(U_i)^+$ and $p_i(U_i)^-$. Here we set $p_i(U_i)^-$ to be the pseudoconvex side of $p_i(U_i \cap X)$. By the extension theorem for the solutions of tangential Cauchy-Riemann equations $p_j p_i^{-1}$ extends to a holomorphic map p_{ji} from an open set of the form $p_i(U_i \cap U_j) \cap B_i$, where $B_i \subset p_i(U_i)^-$ and $\overline{B}_i \supset p_i(U_i \cap U_j \cap X)$. (cf. Theorem 2.6.13 in [7]). Clearly the values are contained in $p_j(U_j)^-$, and we can patch the neighbourhoods of $p_i(U_i \cap X)$ in $p_i(U_i)$ via p_{ji} and obtain a complex manifold with boundary W' whose boundary is isomorphic to X. Replacing W' by a smaller neighbourhood of X if necessary, we may assume that W' is realized in $\mathbb{C}^N$, since we may apply the extension theorem to the

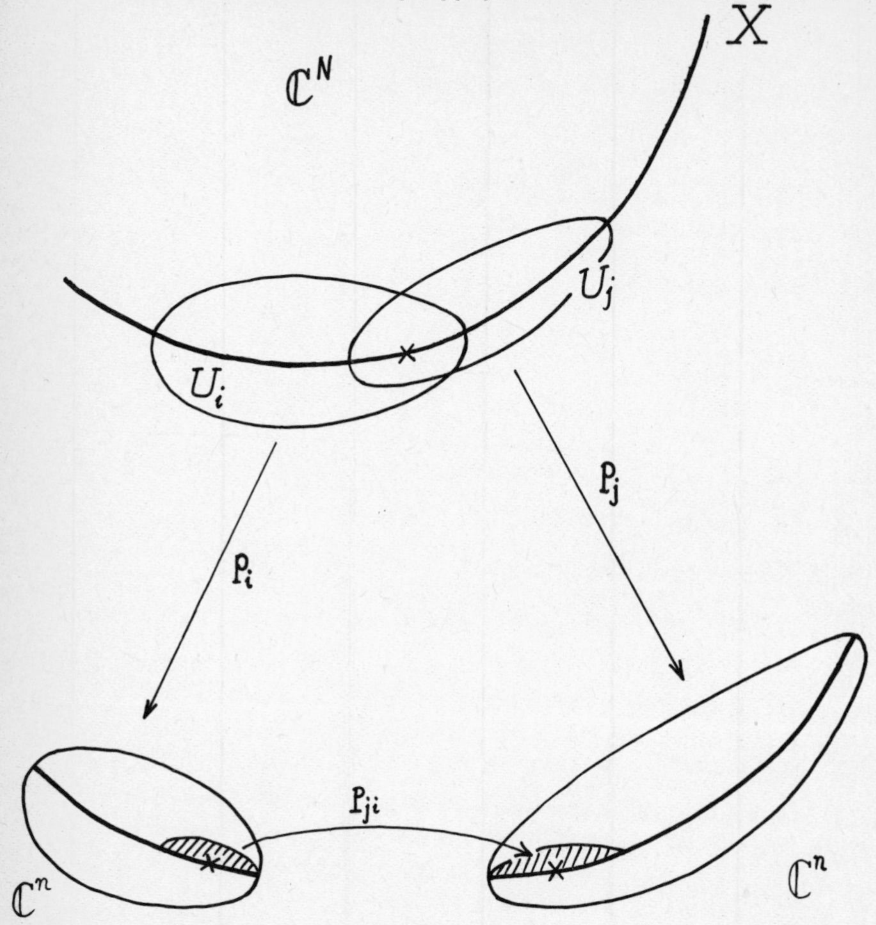

coordinate functions of $\mathbb{C}^N$ restricted to X. Then, in virtue of the extension theorem for analytic sets in $\mathbb{C}^N$, we obtain an analytic subvariety W which satisfies the requirement.

<u>Remark.</u> If SingW is empty, then W is a Stein manifold and is a holomorphic neighbourhood retract. By Kohn [8] the retraction can be chosen to be differentiable up to the boundary. Hence by approximating W by a closed submanifold M in $\mathbb{C}^N$, W is realized as an s.p.c. domain in M.

For the proof of Theorem 2, we need more than the existence of the analytic set W, i.e., we need a complex manifold with boundary $\overline{M}$ which has X as its boundary. The existence of such $\overline{M}$ is assured by Hironaka's desingularization theorem [4]. In fact, it is shown in [4] that by a finite succession of blowing ups we obtain from W a complex manifold M which is biholomorphically equivalent to W near the boundary. We may assume that M has a Kähler metric with smooth boundary values. (cf. [5]).

Notations

$$C^{0,q}(\overline{M}) := \{C^\infty\ (0,q)\text{-forms on } \overline{M}\},$$

$$C_0^{0,q}(\overline{M}) := \{f \in C^{0,q}(\overline{M})\ ;\ f|_X = 0\},$$

$$Z^{0,q}(\overline{M}) := \{f \in C^{0,q}(\overline{M})\ ;\ \bar\partial f = 0\},$$

$$B^{0,q}(\overline{M}) := \bar\partial C^{0,q-1}(\overline{M}),$$

$$H^{0,q}(\overline{M}) := Z^{0,q}(\overline{M})/B^{0,q}(\overline{M}),$$

$$Z_0^{0,q}(\overline{M}) := Z^{0,q}(\overline{M}) \cap C_0^{0,q}(\overline{M}),$$

$$B_0^{0,q}(\overline{M}) := \bar\partial C_0^{0,q-1}(\overline{M}) \cap C_0^{0,q}(\overline{M}),$$

$$H_0^{0,q}(\overline{M}) := Z_0^{0,q}(\overline{M})/B_0^{0,q}(\overline{M}),$$

$$C^{0,q}(X) := \{C^\infty \text{ sections of } \overset{q}{\wedge} T_X'^{*} \longrightarrow X\},$$

$B^{0,q}(X) := \left\{ f \in C^{0,q}(X);\ \text{there exists a } g \in C^{0,q-1}(X) \text{ such that } f = \bar{\partial}_b g := \overset{q}{\wedge} T_X'^{*}\text{-part of } dg \right\},$

$Z^{0,q}(X) := \left\{ f \in C^{0,q}(X);\ \bar{\partial}_b f = 0 \right\},$

$H^{0,q}(X) := Z^{0,q}(X)/B^{0,q}(X).$

By [9] Proposition 6.6, the elements of $H_0^{0,q}(\overline{M})$ are represented by harmonic forms for $q \leq n-1$. Based on this fact, Grauert and Riemenschneider proved that $H_0^{0,q}(\overline{M}) = 0$ for $q \leq n-1$ (see [3]). Thus, from the long exact sequence

$$\cdots\cdots \longrightarrow H^{0,q-1}(X) \longrightarrow H_0^{0,q}(\overline{M}) \longrightarrow H^{0,q}(\overline{M}) \longrightarrow H^{0,q}(X) \longrightarrow H_0^{0,q+1}(\overline{M}) \longrightarrow ;$$

we have $H^{0,q}(\overline{M}) \cong H^{0,q}(X)$, for $q < n-1$.

<u>Proposition 6.</u> Under the above situation,

$$H^{0,q}(\overline{M}) \cong H^{0,q}(M), \quad \text{for } q \geq 1.$$

Proof. Let $\varphi : M \longrightarrow \mathbb{R}$ be a plurisubharmonic function C^∞ up to the boundary such that $\varphi = 0$ on X and $d\varphi \neq 0$ on X. We put $\overline{M}_{-\varepsilon} := \left\{ p \in M;\ \varphi(p) \leq -\varepsilon \right\}$, for $\varepsilon > 0$. For sufficiently small ε , $\partial\overline{M}_{-\varepsilon}$ is also s.p.c. and $\partial\bar{\partial}\varphi > 0$ on $\overline{M} \setminus M_{-\varepsilon}$. Then, harmonic representation theorem for $H^{0,q}(\overline{M})$ and $H^{0,q}(\overline{M}_{-\varepsilon})$ (cf. [8]), and Hörmander's isomorphism theorem ([6] , Theorem 3.4.6 and Theorem 3.4.7.) imply that the natural homomorphism $r_\varepsilon : H^{0,q}(\overline{M}) \longrightarrow H^{0,q}(\overline{M}_{-\varepsilon})$ is an isomorphism for $q \geq 1$. Hörmander's theorem (Theorem 3.4.9 in [6]) implies that $H^{0,q}(M) \cong H^{0,q}(\overline{M}_{-\varepsilon})$ for $q \geq 1$. Therefore, $H^{0,q}(\overline{M}) \cong H^{0,q}(M)$ for $q \geq 1$.

<u>Definition</u> A family of compact s.p.c. manifolds $\{X_t\}_{t\in T}$ is called a differentiable family if there exist a CR manifold $\mathfrak{X}$, a C^∞ manifold T and a proper surjective smooth map $\pi: \mathfrak{X} \longrightarrow T$ such that $(X_t, T'_{X_t}) \cong (\pi^{-1}(t), T'_{\mathfrak{X}} \cap (T_{\pi^{-1}(t)} \otimes \mathbb{C}))$.

<u>Theorem</u> (Tanaka's stability theorem, cf. Theorem 9.4 in [14]) Let $\{X_t\}_{t\in T}$ be a differentiable family of s.p.c. manifolds satisfying the following conditions.

(i) $\dim_{\mathbb{R}} X_t \geq 5$,

(ii) $\dim H^{0,1}(X_t)$ does not depend on t.

Then, for any $t_0 \in T$ and holomorphic embedding $F_0: X_{t_0} \hookrightarrow \mathbb{C}^N$, there exist a neighbourhood $T' \ni t_0$ and a C^∞ map $F: \bigcup_{t\in T'} X_t \longrightarrow \mathbb{C}^N$ such that $F|_{X_{t_0}} = F_0$ and that $F|_{X_t}$ are holomorphic embeddings for every $t \in T'$.

<u>Sketch of the proof</u>: Fix a Riemannian metric on $\mathfrak{X}$. From our conditions (i) and (ii), the Green operator G_t of the Laplacian on $C^{0,1}(X_t)$ differentiablly depends on the parameter t, i.e., for any differntiable family of (0,1)-forms $\varphi_t \in C^{0,1}(X_t)$, $G_t\varphi_t$ is also differntiable with respect to the parameter on $\mathfrak{X}$. Now express F_0 as $(f_1,\dots,f_N)$ and put $f_{i,t}$ to be the harmonic part of the pull-back of f_i by a diffeomorphism $j_t: X_t \xrightarrow{\sim} X_{t_0}$ which depend differentiablly on t. Here t runs through a simply connected neighbourhood of t_0 so that j_t's are well defined. Then, $\bar\partial_b f_{i,t} = 0$ and $f_{i,t}$ depends differentiably on t, since $f_{i,t} = j_t^* f_i - \bar\partial_{b,t}^* G_t \bar\partial_b j_t^* f_i$, where $\bar\partial_{b,t}^*$ denotes the adjoint of $\bar\partial_b$ on X_t.

<u>Remark</u> Without the condition (ii), G_t may not depend differentiably on t. Nevertheless, a differentiable family of holomorphic embeddings of $\{X_t\}_{t\in T}$ <u>seems to</u> exist.

§3. Bumping Family of s.p.c. Manifolds

Let X be an s.p.c. manifold of dimension $2n-1 \geqq 5$, let $x \in X$, and let $B(r)$ be the ball $\{(z_1,\ldots,z_n) \in \mathbb{C}^n; \sum |z_k|^2 < r\}$. By Cor. 2, there exist a neighbourhood $U \ni x$ and an embedding $F: U \hookrightarrow B(1)$ by holomorphic functions so that $F(U)$ is defined by an equation $\varphi = 0$ on $B(1/2)$, where φ is C^∞, $d\varphi \neq 0$ and $\partial\bar\partial\varphi > 0$ on $B(1/2)$. Let ρ be a nonnegative C^∞ function on $B(1/2)$ such that $\operatorname{supp}\rho \subset B(1/4)$ and $\rho = 1$ on $B(1/8)$. Then, for sufficiently small $\varepsilon > 0$, we have $\partial\bar\partial(\varphi - t\rho) > 0$ and $d(\varphi - t\rho) \neq 0$ on a neighbourhood of $F(U) \cap B(1/2)$ for any $t \in [-\varepsilon, \varepsilon]$. Choose a C^∞ family of embeddings $F: U \times [-\varepsilon, \varepsilon] \longrightarrow B(1)$ such that

(1) $\qquad F(x,t) = F(x)$ for $x \in F^{-1}(B(1) \setminus B(1/4))$ and $t \in [-\varepsilon, \varepsilon]$,

(2) $\qquad F(U \times \{t\}) \cap B(1/2) = \{\varphi - t\rho = 0\}$.

F induces a CR structure $T'_{\mathfrak{X}}$ on $\mathfrak{X} := X \times (-\varepsilon, \varepsilon)$ as follows. Let $p: X \times (-\varepsilon, \varepsilon) \longrightarrow X$ and $\pi: X \times (-\varepsilon, \varepsilon) \longrightarrow (-\varepsilon, \varepsilon)$ be the projections. We put $T'_{\mathfrak{X},(x,t)} := p^* T'_{X,x}$ for $x \in X \setminus U$ and put $T'_{\mathfrak{X},(x,t)} := F_t^{-1}{}_*(T'_{F(U \times \{t\}),F(x,t)})$ for $x \in U$, where we put $F_t := F(\ ,t)$. Then we have a differentiable family $\{X_t\}$ of s.p.c. manifolds defined by $(\mathfrak{X}, T'_{\mathfrak{X}}$ and $\pi)$. Let us call it <u>a bumping family of X with front U</u>. Note that it depends on the choices of F, φ and ρ.

Let $\overline{M}_0$ be a complex manifold with boundary such that the boundary is isomorphic to X. Then, by an extension theorem there exist a neighbourhood $V \supset F(U)$ in $B(1)$ and a holomorphic embedding $\iota: V \cap \{\varphi \leqq 0\} \hookrightarrow \overline{M}_0$ such that $\iota = F^{-1}$ on $F(U)$. Thus, for sufficiently small $t > 0$ we can patch $\overline{M}_0$ and $V \cap \{\varphi - t\rho \leqq 0\}$ via the identification $\iota(V \cap \{\varphi \leqq 0\}) \cong V \cap \{\varphi \leqq 0\}$ and obtain a complex manifolds with boundary $\overline{M}_t$ such that $\partial\overline{M}_t \cong X_t$. For nonpositive values of t, we put $\overline{M}_t := \overline{M}_0 \setminus \{\varphi - t\rho > 0\}$. Then $\partial\overline{M}_t \cong X_t$ for t sufficiently close to 0. Choose $\overline{M}_0$ to be $\overline{M}$ constructed in §2. Let $\pi : \overline{M}_0 \longrightarrow \overline{W}$ be the associated proper holomorphic map and let A be the maximal compact analytic subset, i.e., the nowhere discrete compact analytic subset of M_0 such that $\pi|_{M_0 \setminus A}$ is one-to-one. Then we have

$$H^q(M_0, \mathcal{O}_{M_0}) \cong H^q(A, \mathcal{O}_{M_0}|_A) \quad \text{for } q \geq 1$$

(cf. Narasimhan [12]). From the definition of M_t, M_t is holomorphically convex for sufficiently small $|t|$ and contains A as the maximal compact analytic subset (we have used Grauert's theorem [2] here). Hence we can assume that for any M_t with $|t| < \varepsilon$, $H^q(M_0, \mathcal{O}_{M_0}) \cong H^q(M_t, \mathcal{O}_{M_t})$ for $q \geq 1$. Thus, from Prop. 6 and preceeding remark we obtain

<u>Proposition 7.</u> Under the above situation, there exists an $\varepsilon > 0$ such that

$$H^{0,1}(X_t) \cong H^{0,1}(X), \quad \text{for } |t| < \varepsilon.$$

§4. <u>Proof of Theorem 2</u>

We shall now prove every compact s.p.c. manifold of dimension ≥ 5 is realizable as a hypersurface of a complex manifold. Let the notations be as before, and embed X as a CR submanifold of $\mathbb{C}^N$. Take a finite covering $\{V_i\}_{i=1}^m$ of X by open sets of $\mathbb{C}^N$ and choose affine coordinates $(z_{i1}, \ldots, z_{iN})$ so that the projections $p_i : (z_{i1}, \ldots, z_{iN}) \longrightarrow (z_{i1}, \ldots, z_{in})$ induce holomorphic embeddings of $\overline{V_i \cap X}$ into $\overline{B(1)} \subset \mathbb{C}^n$ (an embedding of a n.b.d. of $\overline{V_i \cap X}$), We may assume harmlessly that $\bigcup p_i^{-1}(B(1/8)) \supset X$. Let $\{X_t\}_{|t|<\varepsilon}$ be a bumping family of X with front $V_1 \cap X$.*) Then, by Prop.7, we can apply Tanaka's stability theorem to our family and obtain a differentiable family of holomorphic embeddings $F_{1,t}: X_t \hookrightarrow \mathbb{C}^N$ for $|t| < \delta$ with $0 < \delta < \varepsilon$. Thus we can choose a $t_1 > 0$ such that $\bigcup p_i^{-1}(B(1/8)) \cap V_i \supset F_{1,t_1}(X_{t_1})$ and that p_i induce embeddings of neighbourhoods of $\overline{V_i \cap F_{1,t_1}(X_{t_1})}$. Let $\{\overline{M}_t\}_{|t|<\varepsilon}$ be the family of complex manifolds associated to $\{X_t\}_{|t|<\varepsilon}$, which we constructed in §3. Then we have a natural inclusion $X \subset \overline{M}_{t_1}$. Now inductively we define F_k so that we have a CR manifold $X_{t_1 \ldots t_k}$ as a member of a bumping family of $X_{t_1 \ldots t_{k-1}}$ with front $F_{k-1,t_{k-1}}^{-1}(V_k \cap F_{k-1,t_{k-1}}(X_{t_1 \ldots t_{k-1}}))$. Here we take $t_1 \gg t_2 \gg \ldots \gg t_k > 0$. Let $M_{t_1 \ldots t_m}$ be the complex manifold with boundary corresponding to $X_{t_1 \ldots t_m}$. Clearly, $X \subset M_{t_1 \ldots t_m}$.

Q.E.D.

*) To be precise we should choose the covering $\{V_i\}$ finer if necessary.

REFERENCES

[1] Boutet de Monvel, L.,
Intégration des équations de Cauchy-Riemann induites formelles, Séminaire Goulaouic-Lions-Schwartz 1974-75.

[2] Grauert, H.,
On Levi's problem and the imbedding of real analytic manifolds, Ann. of Math. 68, 460-472 (1958).

[3] Grauert, H. and Riemenschneider, O.,
Kählersche Mannigfaltigkeiten mit hyper-q-konvexem Rand, Problems in Analysis, N.J., Princeton Univ. Press 1970. 61-79.

[4] Hironaka., H.
Resolution of singularities of an algebraic variety over a field of characteristic zero I, II, Ann. of Math. 79 (1964) 109-326.

[5] Hironaka, H., and Rossi, H.,
On the equivalence of imbeddings of exceptional complex spaces, Math. Ann. 156 (1964) 313-333.

[6] Hörmander, L.,
L^2 estimates and existence theorems for the $\bar{\partial}$ operator, Acta Math. 113 (1965), 89-152.

[7] Hörmander, L.,
An introduction to complex analysis in several variables, Van Nostrand, Princeton, N.J., 1966.

[8] Kohn, J.,
Harmonic integrals on strongly pseudo-convex manifolds, I, Ann. of Math. 78 (1963) 112-148.

[9] Kohn, J. and Rossi, H.
On the extension of holomorphic functions from the boundary of a complex manifold, Ann. Math. 81 (1965) 451-472.

[10] Kuranishi, M.,
Strongly pseudoconvex CR structures over small balls Part III. An embedding theorem, Ann. of Math. 116 (1982) 249-330.

[11] Kuranishi, M.,
Application of $\bar{\partial}_b$ to deformation of isolated singularities, Proc. of Symp. in Pure Math. 30, AMS, 1977, 97-106.

[12] Narasimhan, R.,
The Levi problem for complex spaces II, Math. Ann. 146 (1962) 195-216.

[13] Nakano, S. and Ohsawa, T.,
Strongly pseudoconvex manifolds and strongly pseudoconvex domains, to appear in Publ. RIMS, Kyoto Univ.

[14] Tanaka, N.,
A differential geometric study on strongly pseudoconvex manifolds, Lectures in mathematics, Kyoto Univ. 9. 1975.

THE STABILITY AND THE GAUSS MAP OF MINIMAL SURFACES IN $\mathbb{R}^3$

Miyuki Koiso
Department of Mathematics
Osaka University
Toyonaka, Osaka 560, Japan

§0. Introduction.

In this paper we shall be concerned with a certain delicate case for the problem on the stability of minimal surfaces in $\mathbb{R}^3$, referring to the area of the Gaussian image, and give some sufficient conditions for the stability and the instability of minimal surfaces.

Let D be a bounded plane domain whose boundary ∂D is a finite union of piecewise C^∞ curves. Let $\chi:\bar{D} \to \mathbb{R}^3$ be a regular (i.e. immersed) minimal surface. Then χ is a critical point of the area functional for all variations of surfaces which keep their boundary values fixed. When this critical point provides a relative minimum of area for all such variations, we say that it is stable, otherwise unstable.

Set $S = \{(x^1, x^2, x^3) \in \mathbb{R}^3 ; (x^1)^2 + (x^2)^2 + (x^3)^2 = 1\}$ and denote by $\mathfrak{G}:\bar{D} \to S$, the Gauss map of χ. Barbosa and do Carmo [1] gave a sufficient condition for χ to be stable:

Theorem (Barbosa and do Carmo [1]). *If the area of the Gaussian image $\mathfrak{G}(\bar{D})$ of χ is smaller than 2π, then χ is stable.*

This estimate is sharp in the following sense: There are examples of unstable minimal surfaces whose Gaussian image has area larger than 2π and as close to 2π as one pleases.

Then what can we say about the stability of χ in the case when the area of $\mathfrak{G}(\bar{D})$ is exactly 2π? The purpose of this paper is to answer this problem.

In the first two cases, Case 1 and Case 2 mentioned below, χ is always stable (Theorem 1 and Theorem 2).

Case 1. $\mathfrak{G}(\bar{D})$ does not coincide with any hemisphere H of S.

Case 2. $\mathfrak{G}(\bar{D})$ coincides with a hemisphere H and $\mathfrak{G}(\partial D) \neq \partial H$.

In the remaining case, Case 3 (i.e. $\mathfrak{G}(\bar{D}) = H$ and $\mathfrak{G}(\partial D) = \partial H$), let f, g be the factors of the Weierstrass representation of x (cf. §2). By a suitable rotation of the surface in $\mathbb{R}^3$, we may assume that $\mathfrak{G}(\bar{D})$ coincides with the lower hemisphere of $S : H^- = \{(x^1, x^2, x^3) \in S ; x^3 \leq 0\}$. In this situation, g is a holomorphic function of D onto $D_0 = \{w \in \mathbb{C} ; |w| < 1\}$. Here we define a function F in D_0 as follows:

$$F(w) = \sum_{\{\zeta \in D; g(\zeta)=w\}} \frac{g'(\zeta)}{f(\zeta)}.$$

Then F is seen to be holomorphic in D_0 and we can prove the following fact:

Main theorem. Let $x:\bar{D} \to \mathbb{R}^3$ be a regular minimal surface and let $\mathfrak{G}:\bar{D} \to S$ be the Gauss map of x. Suppose that $\mathfrak{G}(\bar{D})$ coincides with the lower hemisphere H^- of S and that $\mathfrak{G}(\partial D) = \partial H$. If

$$\operatorname{Re} F''(0) \neq 0,$$

then x is unstable.

Therefore every minimal surface satisfying the assumption of the above theorem is not physically realized as soap film. It must be interesting that the instability of x is decided only by the values of derivatives of x at a finite number of points which are mapped to the south pole of S by the Gauss map $\mathfrak{G}$. This result is proved by calculating the third variation of area functional.

§1. Notations and terminology.

A minimal surface x in $\mathbb{R}^3$ is a C^2 mapping x from some domain D in the plane into $\mathbb{R}^3$ which is harmonic in D, extends continuously to the closure $\bar{D}$, and satisfies $x_\xi \cdot x_\eta = 0$, $|x_\xi| = |x_\eta|$ in D (where $\zeta = \xi + \sqrt{-1}\,\eta$ is the variable in the parameter domain). A branch point of x is some point $\zeta \in \bar{D}$ where $x_\xi = x_\eta = 0$. Branch points are the only possible singularities of minimal surfaces.

In the following the parameter domain D is supposed to be a bounded domain whose boundary is a finite union of piecewise C^∞ curves. And we shall be concerned only with regular minimal surfaces which can be extended as minimal surfaces across ∂D, where "regular" means that the surface considered has no branch points on $\bar{D}$.

If $\mathfrak{y}:\bar{D} \to \mathbb{R}^3$ is of piecewise C^1-class on $\bar{D}$, then the area functional is defined as

$$A(\mathfrak{y}) = \iint_D |\mathfrak{y}_\xi \times \mathfrak{y}_\eta|\, d\xi d\eta = \iint_D \sqrt{E_\mathfrak{y} G_\mathfrak{y} - F_\mathfrak{y}^2}\, d\xi d\eta,$$

where $E_\mathfrak{y} = \mathfrak{y}_\xi^2$, $F_\mathfrak{y} = \mathfrak{y}_\xi \cdot \mathfrak{y}_\eta$, and $G_\mathfrak{y} = \mathfrak{y}_\eta^2$.

We give here the rigorous definitions for the stability and the instability of minimal surfaces. Let us introduce a function space $C_0^{2'}(\bar{D})$ as

$$C_0^{2'}(\bar{D}) = \{u:\bar{D} \to \mathbb{R}\ ;\ u \text{ is a piecewise } C^2 \text{ function with } u|_{\partial D}=0.\}.$$

For each smooth family $v(\varepsilon) \in C_0^{2'}(\bar{D})$ (ε runs in an interval containing zero and "smooth" means that $v(\varepsilon)$ is smooth with respect to ε and the derivatives are contained in $C_0^{2'}(\bar{D})$) with $v(0) = 0$ and $[\partial v(\varepsilon)/\partial \varepsilon]_{\varepsilon=0} \neq 0$, we consider the normal variation of $\mathfrak{x} : \mathfrak{x} + v(\varepsilon)\mathfrak{G}$, where the Gauss map $\mathfrak{G}$ is identified with the unit normal vector field of $\mathfrak{x}$.

<u>Definition</u>. (i) A minimal surface $\mathfrak{x}$ is said to be <u>stable</u> if for each smooth family $v(\varepsilon) \in C_0^{2'}(\bar{D})$ with $v(0) = 0$ and $[\partial v(\varepsilon)/\partial\varepsilon]_{\varepsilon=0} \neq 0$, there exists some $\varepsilon_0 > 0$ such that

$$A(\mathfrak{x}) \le A(\mathfrak{x} + v(\varepsilon)\mathfrak{G})$$

holds for every ε, $|\varepsilon| < \varepsilon_0$.

(ii) A minimal surface $\mathfrak{x}$ is said to be <u>unstable</u> if $\mathfrak{x}$ is not stable, that is, there exists some smooth family $v(\varepsilon) \in C_0^{2'}(\bar{D})$ with $v(0) = 0$ and $[\partial v(\varepsilon)/\partial\varepsilon]_{\varepsilon=0} \neq 0$, such that for each $\varepsilon_0 > 0$,

$$A(\mathfrak{x}) > A(\mathfrak{x} + v(\varepsilon)\mathfrak{G})$$

holds for some ε, $|\varepsilon| < \varepsilon_0$.

The notations and symbols below will be used throughout this paper without particular mentions:

$\zeta = \xi + \sqrt{-1}\,\eta$ $(\xi, \eta \in \mathbb{R})$: the variable in the parameter domain;
$\Delta = \partial^2/\partial\xi^2 + \partial^2/\partial\eta^2$: the Laplacian on the parameter domain;

∂D : the boundary of D;

$\overline{D}$: the closure of D;

D_0 : the unit open disk;

S : the unit sphere in $\mathbb{R}^3$;

H : a (closed) hemisphere of S;

H^- : $\{(x^1, x^2, x^3) \in S ; x^3 \leq 0\}$: the lower (closed) hemisphere of S;

$\mathfrak{X}:\overline{D} \to \mathbb{R}^3$: a regular minimal surface which can be extended as a minimal surface across ∂D;

$E\,d\xi^2 + 2F\,d\xi d\eta + G\,d\eta^2$: the first fundamental form of $\mathfrak{X}$;

$L\,d\xi^2 + 2M\,d\xi d\eta + N\,d\eta^2$: the second fundamental form of $\mathfrak{X}$;

$W = \sqrt{EG - F^2}$: the area element of $\mathfrak{X}$;

$\mathfrak{G} = \mathfrak{X}_\xi \times \mathfrak{X}_\eta / |\mathfrak{X}_\xi \times \mathfrak{X}_\eta| : \overline{D} \to S$: the Gauss map (sometimes identified with the unit normal vector field) of $\mathfrak{X}$;

K : the Gaussian curvature of $\mathfrak{X}$;

Since our surface $\mathfrak{X}$ is minimal, it follows that

$$W = E = G, \qquad F = 0,$$

and

$$L + N = \mathfrak{G}\cdot(\mathfrak{X}_{\xi\xi} + \mathfrak{X}_{\eta\eta}) = 0.$$

Therefore

$$K = \frac{LN - M^2}{EG - F^2} = -\frac{L^2 + M^2}{W^2}. \tag{1}$$

Remark 1. For each subdomain $\tilde{D}$ of the parameter domain, we denote the image of $\tilde{D}$ under $\mathfrak{G}$ by $\mathfrak{G}(\tilde{D}) = \{\mathfrak{G}(\zeta) ; \zeta \in \tilde{D}\}$. Although $\mathfrak{G}$ is a complex analytic mapping of $\tilde{D}$, we regard $\mathfrak{G}(\tilde{D})$ as a mere subset of S, ignoring its number of sheets.

§2. The Weierstrass representation and the second fundamental form.

In this section we recall the Weierstrass representation of minimal surfaces and derive a certain important relation between the factors of the representation and the second fundamental form of the surface. The facts mentioned in this section will be used effectively in §6 to investigate the stability of a certain kind of minimal surfaces.

Since $x = (x^1, x^2, x^3): \bar{D} \to \mathbb{R}^3$ is a minimal surface, each of the functions

$$\phi_j = \frac{\partial x^j}{\partial \xi} - \sqrt{-1}\,\frac{\partial x^j}{\partial \eta}, \qquad j = 1, 2, 3, \tag{2}$$

is holomorphic in D. Let us introduce two functions with Enneper-Weierstrass:

$$f = \phi_1 - \sqrt{-1}\,\phi_2, \qquad g = \frac{\phi_3}{\phi_1 - \sqrt{-1}\,\phi_2}. \tag{3}$$

Then f is holomorphic and g is meromorphic in D. Moreover, $f(\zeta) = 0$ if and only if ζ is a branch point of x. Hence, for our regular minimal surface x, $f(\zeta)$ vanishes nowhere, and moreover, $|f(\zeta)| > \delta$ for some $\delta > 0$ on D. Since ζ is an isothermal parameter of x, it follows that

$$\phi_1 = \frac{1}{2} f(1-g^2), \qquad \phi_2 = \frac{\sqrt{-1}}{2} f(1+g^2), \qquad \phi_3 = fg. \tag{4}$$

Therefore if $\zeta_0 \in D$,

$$\begin{pmatrix} x^1(\zeta) \\ \\ x^2(\zeta) \\ \\ x^3(\zeta) \end{pmatrix} = \begin{pmatrix} \mathrm{Re} \displaystyle\int_{\zeta_0}^{\zeta} \frac{1}{2} f(1 - g^2)\, d\zeta \\ \\ \mathrm{Re} \displaystyle\int_{\zeta_0}^{\zeta} \frac{\sqrt{-1}}{2} f(1 + g^2)\, d\zeta \\ \\ \mathrm{Re} \displaystyle\int_{\zeta_0}^{\zeta} fg\, d\zeta \end{pmatrix} + \begin{pmatrix} x^1(\zeta_0) \\ \\ x^2(\zeta_0) \\ \\ x^3(\zeta_0) \end{pmatrix}.$$

This representation is called the Weierstrass representation of the minimal surface x. Let us call f, g the factors of the Weierstrass representation (or the W-factors) of x.

From the equations (2) and (4) we derive

$$W = |x_\xi|^2 = |x_\eta|^2 = \frac{1}{2} \sum_{j=1}^{3} |\phi_j|^2 = \left\{ \frac{|f|(1 + |g|^2)}{2} \right\}^2. \tag{5}$$

Now, by some calculations, we observe

$$\mathfrak{G} = \left(\frac{2\ \mathrm{Re}\ g}{|g|^2 + 1}, \frac{2\ \mathrm{Im}\ g}{|g|^2 + 1}, \frac{|g|^2 - 1}{|g|^2 + 1}\right). \tag{6}$$

Consequently g coincides with the composition $P \circ \mathfrak{G}$ of the Gauss map $\mathfrak{G}$ with the stereographic projection P from the point $(0, 0, 1)$ onto the (x^1, x^2)-plane.

The following proposition will give some information about the geometrical meaning of the holomorphic function f.

Proposition 1. Let $x : \bar{D} \to \mathbb{R}^3$ be a minimal surface and let f, g be the W-factors of x. Denote the second fundamental form of x by $\alpha = L\, d\xi^2 + 2M\, d\xi d\eta + N\, d\eta^2$. Then

$$L - \sqrt{-1}\, M = -\, fg', \tag{7}$$

$$\alpha = -\,(1/2) fg'\, d\zeta^2 - (1/2)\overline{fg'}\, d\bar{\zeta}^2,$$

where $'$ means the derivative of a holomorphic function, $d\zeta = d\xi + \sqrt{-1}\, d\eta$, and $d\bar{\zeta} = d\xi - \sqrt{-1}\, d\eta$.

Proof. Set $f^{(1)} = \mathrm{Re}\, f$, $f^{(2)} = \mathrm{Im}\, f$, $g^{(1)} = \mathrm{Re}\, g$, and $g^{(2)} = \mathrm{Im}\, g$. By the equation (6), we see

$$\begin{aligned}\mathfrak{G}_\xi = \Big(& 2\big\{-g^{(1)}_\xi (g^{(1)})^2 + g^{(1)}_\xi (g^{(2)})^2 - 2g^{(2)}_\xi g^{(1)} g^{(2)} + g^{(1)}_\xi\big\}\big/(|g|^2+1)^2, \\ & 2\big\{g^{(2)}_\xi (g^{(1)})^2 - g^{(2)}_\xi (g^{(2)})^2 - 2g^{(1)}_\xi g^{(1)} g^{(2)} + g^{(2)}_\xi\big\}\big/(|g|^2+1)^2, \\ & 4(g^{(1)}_\xi\, g^{(1)} + g^{(2)}_\xi\, g^{(2)})\big/(|g|^2 + 1)^2\Big).\end{aligned} \tag{8}$$

On the other hand,

$$\begin{aligned} L - \sqrt{-1}\, M &= -\,\mathfrak{G}_\xi \cdot x_\xi - \sqrt{-1}\,(-\,\mathfrak{G}_\xi \cdot x_\eta) \\ &= -\,\mathfrak{G}_\xi \cdot (\phi_1, \phi_2, \phi_3) \\ &= -\,\mathfrak{G}_\xi \cdot \big(f(1-g^2)/2,\ \sqrt{-1}\, f(1+g^2)/2,\ fg\big). \end{aligned} \tag{9}$$

By some calculation using (8) and (9), we obtain

$$L - \sqrt{-1}\, M = -\, fg'.$$

Since x is minimal, $N = -L$ (§1), which leads us to the second formula of this proposition with the help of equation (7). Q.E.D.

Although the following lemma is well known, we contain a brief sketch of its proof by using the above proposition.

Lemma 1.

$$K = -\left\{\frac{4|g'|}{|f|(1+|g|^2)^2}\right\}^2 . \tag{10}$$

Therefore K is non-positive. Moreover, K can have only isolated zeros unless the locus of χ lies entirely in a plane.

Proof. By using (1), (5), and Proposition 1, we obtain the equation (10). Since g is meromorphic, g' can have only isolated zeros unless g' is identically zero. Moreover, $g' \equiv 0$ if and only if $\mathbb{G}$ is constant by virtue of the equation $g = P\circ\mathbb{G}$. Q.E.D.

§3. The variations of area and the eigenvalue problem associated with area.

Let $v(\varepsilon)$ be a smooth family in $C_0^{2'}(\bar{D})$ with $v(0) = 0$ and $[\partial v(\varepsilon)/\partial\varepsilon]_{\varepsilon=0} = u \neq 0$. With the aid of the minimality of χ, the first and the second variations of area functional for a normal variation $\chi + v(\varepsilon)\mathbb{G}$ are given by the following formulae (cf. Beeson [2]):

$$\frac{d}{d\varepsilon} A(\chi + v(\varepsilon)\mathbb{G})\big|_{\varepsilon=0} = 0,$$

$$\frac{d^2}{d\varepsilon^2} A(\chi + v(\varepsilon)\mathbb{G})\big|_{\varepsilon=0} = \iint_D u(-\Delta u + 2KWu)\, d\xi d\eta.$$

Therefore, as for the first and the second variations of area, it is sufficient to consider only variations formed as $\chi + \varepsilon u\mathbb{G}$ $(\varepsilon \in \mathbb{R},\ u \in C_0^{2'}(\bar{D})$ and $u \neq 0)$ which we shall call "variation u". Then

$$I^{(1)}(u) = \frac{d}{d\varepsilon} A(\chi + \varepsilon u\mathbb{G})\big|_{\varepsilon=0} = \frac{d}{d\varepsilon} A(\chi + v(\varepsilon)\mathbb{G})\big|_{\varepsilon=0} = 0, \tag{11}$$

$$\begin{aligned} I^{(2)}(u) &= \frac{d^2}{d\varepsilon^2} A(\chi + \varepsilon u\mathbb{G})\big|_{\varepsilon=0} \\ &= \frac{d^2}{d\varepsilon^2} A(\chi + v(\varepsilon)\mathbb{G})\big|_{\varepsilon=0} = \iint_D u(-\Delta u + 2KWu)\, d\xi d\eta. \end{aligned} \tag{12}$$

Moreover the third variation of area functional for variation u is given by

$$I^{(3)}(u) = \frac{d^3}{d\varepsilon^3} A(\mathfrak{X}+\varepsilon u\mathfrak{G})\Big|_{\varepsilon=0} = \iint_D \frac{6u}{W}\left\{L(u_\xi{}^2-u_\eta{}^2)+2Mu_\xi u_\eta\right\} d\xi d\eta \qquad (13)$$

(cf. Nitsche [4, p.93]).

From now on, we assume that the locus of $\mathfrak{X}$ is not contained in a plane. Therefore, by virtue of Lemma 1, K can have only isolated zeros.

Now we consider an eigenvalue problem related to the second variation above. Let $\tilde{D} \subseteq D$ be a subdomain of D such that $\partial\tilde{D}$ is piecewise C^∞. Then we pose the eigenvalue problem:

$$\begin{cases} \Delta u - \lambda KWu = 0 & \text{in } \tilde{D}, \\ u = 0 & \text{on } \partial\tilde{D}. \end{cases} \qquad (14)$$

If we denote by $\lambda_1(\tilde{D})$ the least eigenvalue of the problem (14), then we see almost immediately from Beeson [2] the following

Lemma 2. (i) If $D_1 \subseteq D_2$, then $\lambda_1(D_1) \geq \lambda_1(D_2)$, where the equality holds if and only if $D_1 = D_2$.

(ii) According as $\partial\tilde{D}$ varies smoothly, $\lambda_1(\tilde{D})$ varies continuously.

(iii) $\lambda_1(\tilde{D})$ is equal to the minimum of the Rayleigh quotient:

$$R(u) = \frac{\iint_{\tilde{D}} (-u\Delta u)\, d\xi d\eta}{\iint_{\tilde{D}} (-KW)u^2\, d\xi d\eta}, \qquad u \in C_0^{2'}(\overline{\tilde{D}}),$$

and the equality "$R(u) = \lambda_1(\tilde{D})$" holds if and only if u is a least eigenfunction (i.e. the one associated with the least eigenvalue $\lambda_1(\tilde{D})$).

(iv) Each least eigenfunction has the definite sign. But except them, every eigenfunction changes its sign.

(v) The eigenspace corresponding to the least eigenvalue is 1 - dimensional.

Now set $\tilde{D} = D$ in (14). By using the fact that $\lambda_1(D)$ minimizes $R(u)$ ((iii) of the above lemma), we can derive the relationship between the least eigenvalue of the problem (14) and the stability of the minimal surface:

Lemma 3. (i) If $\lambda_1(D) > 2$, then $I^{(2)}(u) > 0$ for all the variations u. Therefore $\mathfrak{X}$ is stable.

(ii) If $\lambda_1(D) = 2$, then $I^{(2)}(u) \geq 0$ for all such variations u, and $I^{(2)}(u) = 0$ holds if and only if u is a least eigenfunction of (14).

(iii) If $\lambda_1(D) < 2$, then there exists some u such that $I^{(2)}(u) < 0$. Therefore χ is unstable.

Proof. Since λ_1 $(= \lambda_1(D))$ minimizes $R(u)$,

$$\lambda_1 \le \frac{\iint_D (-u\Delta u)\, d\xi d\eta}{\iint_D (-KW)u^2\, d\xi d\eta}$$

for every variation u. By using the fact that $-KW$ is non-negative (Lemma 1), we see

$$\lambda_1 \iint_D (-KW)u^2\, d\xi d\eta \le \iint_D (-u\Delta u)\, d\xi d\eta.$$

Therefore

$$\begin{aligned} I^{(2)}(u) &= \iint_D u(-\Delta u + 2KWu)\, d\xi d\eta \\ &\ge (\lambda_1 - 2)\iint_D (-KW)u^2\, d\xi d\eta. \end{aligned}$$

Hence, the assumption $\lambda_1 > 2$ implies $I^{(2)}(u) > 0$ for all u. Moreover, if $\lambda_1 = 2$, $I^{(2)}(u) \ge 0$ for all u, and $I^{(2)}(u) = 0$ if and only if $\lambda_1 = R(u)$ which is just the case in which u is a least eigenfunction by virtue of Lemma 2 (iii). Thus we have proved (i) and (ii).

Suppose that $\lambda_1 < 2$ and that u is a least eigenfunction. Then

$$2 > \lambda_1 = \frac{\iint_D (-u\Delta u)\, d\xi d\eta}{\iint_D (-KW)u^2\, d\xi d\eta}.$$

Therefore

$$\begin{aligned} I^{(2)}(u) &= \iint_D u(-\Delta u + 2KWu)\, d\xi d\eta \\ &= (\lambda_1 - 2)\iint_D (-KW)u^2\, d\xi d\eta \\ &< 0. \end{aligned}$$

Q.E.D.

Remark 2. In the case (ii) of Lemma 3, we cannot so easily arrive at any conclusion about the stability of the minimal surface χ. calculating the third variation is one of the ways to obtain some conclusion. In fact, if, for some u, $I^{(2)}(u) = 0$ and $I^{(3)}(u) \neq 0$, then x is clearly unstable.

§4. Case 1. $\mathbb{G}(\overline{D})$ has area 2π and does not coincide with H.

From now on, we assume that the area of $\mathbb{G}(\overline{D})$ equals exactly 2π. At first, in this section, we are concerned with the case in which $\mathbb{G}(\overline{D})$ does not coincide with any hemisphere H of S.

When we regard S as a Riemannian manifold with the Riemannian metric induced from the Euclidean space $\mathbb{R}^3$, we denote the Laplacian for functions in S by Δ_S. Let Ω be a domain in S. Consider the eigenvalue problem:

$$\begin{cases} \Delta_S v + \lambda v = 0 & \text{in } \Omega, \\ v = 0 & \text{on } \partial\Omega. \end{cases} \tag{15}$$

We denote by $\tilde{\lambda}_1(\Omega)$ the least eigenvalue of this problem.

Lemma 4 (Peetre [5]). *Among all spherical domains with the same area, only the spherical cap minimizes* $\tilde{\lambda}_1$.

Lemma 5 (Barbosa and do Carmo [1]). $\tilde{\lambda}_1(\text{Int } H) = 2$, *where* Int H *stands for the interior of* H.

Lemma 6 (Barbosa and do Carmo [1]). *If* $\tilde{\lambda}_1(\mathbb{G}(D)) > 2$, *then* $I^{(2)}(u) > 0$ *for all* u, *so* x *is stable*.

Now that $\tilde{\lambda}_1(\mathbb{G}(D)) > 2$ for our minimal surface x by Lemmas 4 and 5, Lemma 6 is applicable. Thus we have proved

Theorem 1. Let the image of the Gauss map of a regular minimal surface x *have area* 2π. *Assume that it does not coincide with any hemisphere of* S. *Then* x *is stable*.

§5. Case 2. $\mathbb{G}$ maps $\overline{D}$ onto H and $\mathbb{G}(\partial D) \neq \partial H$.

Theorem 2. Let the image of the Gauss map $\mathfrak{G}$ of a regular minimal surface x coincide with a hemisphere H of S. Suppose that $\mathfrak{G}(\partial D) \neq \partial H$. Then the second variation of area is always positive, and hence x is stable.

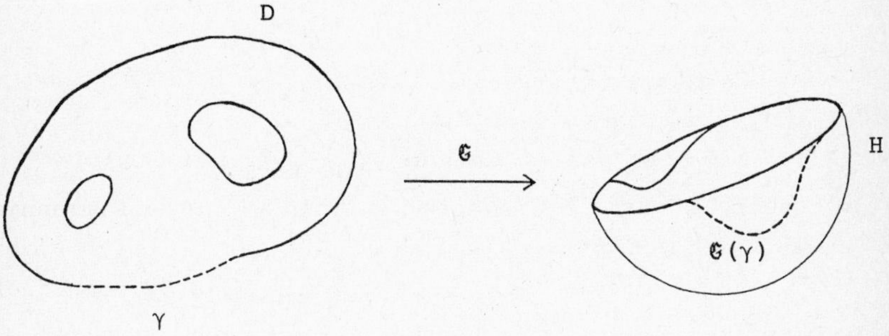

[Fig.1]

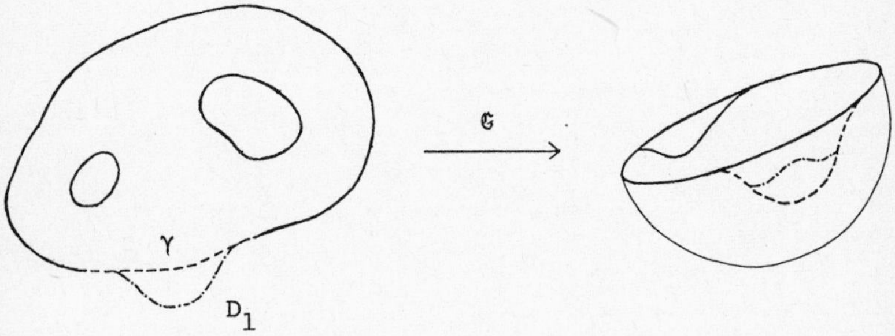

[Fig.2]

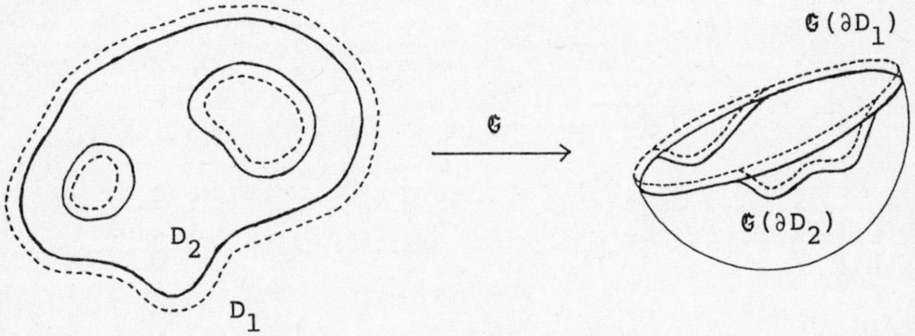

[Fig.3]

<u>Proof</u>. Assume that $I^{(2)}(u) \le 0$ for some u. Then $\lambda_1(D) \le 2$ (Lemma 3). Since $\mathfrak{G}(\partial D) \ne \partial H$, there exists some arc $\gamma \subset \partial D$ such that $\mathfrak{G}(\gamma) \subset \mathrm{Int}\, H$ (Fig.1). By assumption, x can be extended across γ up to some domain $D_1 \supsetneq D$ such that $\mathfrak{G}(\overline{D_1}) = H$, too (Fig.2). Because λ_1 is strictly decreasing (Lemma 2 (i)), $\lambda_1(D_1) < \lambda_1(D) \le 2$. Owing to this fact and the continuity of λ_1 (Lemma 2 (ii)), we can take some domain D_2 whose closure is contained in $\mathrm{Int}\, D_1$ such that $\lambda_1(D_2) < 2$ (Fig.3).

On the other hand, the second W-factor g of $x:\overline{D_1} \to \mathbb{R}^3$ is holomorphic in D_1 and $g(\overline{D_1}) = \overline{D_0} = \{w \in \mathbb{C} ; |w| \le 1\}$. Therefore $g(D_2) \subsetneq D_0$ by the maximum principle, which implies that $\mathfrak{G}(D_2) \subsetneq \mathrm{Int}\, H$ (Fig. 3). Since $\tilde{\lambda}_1$ is decreasing, $\tilde{\lambda}_1(\mathfrak{G}(D_2)) > 2$ (Lemma 5). Thus, by applying Lemmas 3 and 6 to the minimal surface $x|_{D_2}$, we obtain $\lambda_1(D_2) > 2$. This is a contradiction. Q.E.D.

§6. <u>Case 3. $\mathfrak{G}$ maps $\overline{D}$ onto H and $\mathfrak{G}(\partial D) = \partial H$.</u>

Finally we consider the case in which $\mathfrak{G}(\overline{D})$ coincides with a hemisphere H of S and $\mathfrak{G}(\partial D) = \partial H$. By a suitable rotation of the surface in $\mathbb{R}^3$, we may assume that H coincides with the lower hemisphere H^- $= \{(x^1, x^2, x^3) \in S ; x^3 \le 0\}$. Then $g = P\circ\mathfrak{G}$ (see §2) maps D onto the unit open disk D_0.

<u>Lemma 7</u>. <u>Suppose that</u> $\mathfrak{G}(\overline{D}) = H^-$ <u>and that</u> $\mathfrak{G}(\partial D) = \partial H^-$. <u>Then</u> $g|_D : D \to D_0$ <u>is a finite-sheeted branched covering and the number of branch points of</u> g <u>is finite</u>.

(For the notion of (global) branched covering, cf. Gunning [3, pp. 220-221].)

<u>Proof</u>. Because x can be extended as a minimal surface across ∂D, g is holomorphic on $\overline{D}$. From this fact and the fact that $g(\partial D) = \partial D_0$, the lemma holds immediately.

<u>Lemma 8</u>. <u>It holds that</u> $\lambda_1(D) = 2$. <u>And the eigenspace corresponding to the least eigenvalue of the problem (14) is given by</u>

$$E_1 = \left\{ a v_0 \circ \mathfrak{G} \; ; \; v_0((x^1, x^2, x^3)) \equiv x^3, \; a \in \mathbb{R} \right\}.$$

Proof. Put $u_0 = v_0 \circ \mathfrak{G}$ with $v_0((x^1, x^2, x^3)) \equiv x^3$. Then by (6)

$$u_0(\zeta) \equiv \frac{|g(\zeta)|^2 - 1}{|g(\zeta)|^2 + 1}. \tag{16}$$

From some easy calculations, we derive

$$\Delta u_0 = \frac{8|g'|^2(1 - |g|^2)}{(|g|^2 + 1)^3}. \tag{17}$$

From (5), (10), (16), and (17), we conclude that

$$\Delta u_0 - 2KWu_0 = 0.$$

Moreover, $u_0 < 0$ on D and $u_0|_{\partial D} = 0$. Therefore u_0 is one of the least eigenfunctions of the problem (14) (Lemma 2 (iv)), and the least eigenvalue $\lambda_1(D)$ equals 2. Since the eigenspace corresponding to λ_1 is 1-dimensional (Lemma 2 (v)), the proof is completed.

From Lemmas 3 and 8, we observe

Lemma 9. $I^{(2)}(u) \geq 0$ *for all* u. *The equality holds if and only if* $u \in E_1$.

In view of Remark 2 and the above lemma we shall investigate the third variation of area for $u \in E_1$.

Lemma 10. *If* u *belongs to* E_1, *then*

$$I^{(3)}(u) = \mathrm{Re} \iint_D \frac{384a^3(1 - |g|^2)|g|^4|g'|^2}{(1 + |g|^2)^7} \cdot \frac{g'}{fg^2}\, d\xi d\eta, \tag{18}$$

where a *is a real constant determined by the choice of* u, i.e. $u = av_0$ (*cf.* Lemma 8).

Proof. By virtue of (5), (7), and (13), we see

$$I^{(3)}(u) = \iint_D \frac{-24u}{|f|^2(1 + |g|^2)^2} \cdot \mathrm{Re}\left\{fg'(u_\xi + \sqrt{-1}\, u_\eta)^2\right\} d\xi d\eta. \tag{19}$$

Since $u \in E_1$, u can be written in the form:

$$u = a \cdot \frac{|g|^2 - 1}{|g|^2 + 1}, \qquad a \in \mathbb{R}$$

((16) in the proof of Lemma 8). Therefore, by some calculation, we see

$$(u_\xi + \sqrt{-1}\, u_\eta)^2 = \frac{16a^2 g^2 \overline{g'}^2}{(|g|^2 + 1)^4}. \tag{20}$$

Using (19) and (20), we obtain the equation (18). Q.E.D.

Since $g: D \to D_0$ is a finite-sheeted branched covering (Lemma 7), for each point $\zeta \in D$, there exists some open neighborhood $U \subset D$ of ζ, such that $g(U)$ is open in D_0 and the restriction of g to $U - \{\zeta\}$ is a finite-sheeted covering of $g(U) - \{g(\zeta)\}$. The number of such sheets is called the multiplicity of the point ζ for g.

Let us introduce a single-valued function F in D_0 as

$$F(w) = \sum_{\{\zeta \in D;\, g(\zeta) = w\}} \frac{g'(\zeta)}{f(\zeta)}. \tag{21}$$

The following lemma will play an important rôle in subsequent arguments.

<u>Lemma 11</u>. F <u>is holomorphic in</u> D_0.

<u>Proof</u>. Denote the branch points of g by $\zeta_1, \cdots, \zeta_m$ (Lemma 7). Set $D_1 = D - \{\zeta_1, \cdots, \zeta_m\}$ and $D_2 = D_0 - \{g(\zeta_1), \cdots, g(\zeta_m)\}$.

Let $q \in D_0$, let $g^{-1}(q) = \{p_1, \cdots, p_n\}$ ($p_j \neq p_k$, if $j \neq k$), and let ν_j be the multiplicity of p_j, $j = 1, \cdots, n$. Then there exists a neighborhood $V \subset D_2 \cup \{q\}$ of q such that $g^{-1}(V)$ is the disjoint union of neighborhoods $U_1, \cdots, U_n \subset D$ of the points $p_1, \cdots, p_n$, respectively, and the restriction of g to $U_j - \{p_j\}$ is a ν_j-sheeted covering of $V - \{q\}$, $\nu_j \geq 1$. Define functions F_j in V as follows:

$$F_j(q) = \frac{g'(p_j)}{f(p_j)}.$$

and if $\tilde{q} \in V - \{q\}$,

$$F_j(\tilde{q}) = \sum_{k=1}^{\nu_j} \frac{g'(\tilde{p}_k^j)}{f(\tilde{p}_k^j)},$$

where $g^{-1}(\tilde{q}) \cap U_j = \{\tilde{p}_1^j, \cdots, \tilde{p}_{\nu_j}^j\}$. Then $F|_V = \sum_{j=1}^n F_j$. If we note that $\nu_j > 1$ if and only if $g'(p_j) = 0$, then we can observe that each F_j is continuous in V. Therefore F is continuous in D_0.

When $q \in D_2$, then each restriction $g_j = g|_{U_j}$ has a holomorphic inverse $g_j^{-1}: V \to U_j$. Therefore

$$F = \sum_{j=1}^n F_j = \sum_{j=1}^n \frac{g'}{f} \circ g_j^{-1}$$

is holomorphic in V. Because of the arbitrariness of q, F is holomorphic in D_2. Since F is continuous in D_0 as mentioned above, $g(\zeta_1), \cdots, g(\zeta_m)$ are removable singularities of $F|_{D_2}$, and F is holomorphic in the whole D_0.

Remark 3. F remains invariant under any parameter change of the "minimal surface" X. Namely, let $\tau:\tilde{D} \to D$ be a conformal mapping from some domain $\tilde{D}$ in the plane onto D, then $\tilde{F}$ which is constructed from the minimal surface $\tilde{X} = X \circ \tau$ by (21) coincides exactly with F.

Proposition 2. *If* u *belongs to* E_1, *then*

$$I^{(3)}(u) = \pi a^3 \cdot \mathrm{Re}\,\{F''(0)\}, \tag{22}$$

where a *is a real constant determined by the choice of* u, *i.e.* $u = av_0$ *(cf. Lemma 8)*.

Proof. Let us first prove that

$$I^{(3)}(u) = \mathrm{Re} \iint_{D_0} \frac{384a^3(1 - |w|^2)|w|^4}{(1 + |w|^2)^7} \cdot \frac{F(w)}{w^2}\, dx^1 dx^2, \tag{23}$$

where $w = x^1 + \sqrt{-1}\, x^2$. Set

$$A(\zeta) = \frac{384a^3(1 - |g(\zeta)|^2)|g(\zeta)|^4}{(1 + |g(\zeta)|^2)^7} \cdot \frac{g'(\zeta)}{f(\zeta)\{g(\zeta)\}^2},$$

then

$$I^{(3)}(u) = \mathrm{Re} \iint_D A(\zeta)|g'(\zeta)|^2\, d\xi d\eta$$

by virtue of Lemma 10.

Let D_1 and D_2 be what were defined in the proof of Lemma 11. Let $q \in D_2$ and let $g^{-1}(q) = \{p_1, \cdots, p_n\}$. Then similarly to the proof of Lemma 11, there exists some neighborhood $V \subset D_2$ of q such that $g^{-1}(V)$ is the disjoint union of neighborhoods $U_1, \cdots, U_n \subset D_1$ of the points $p_1, \cdots, p_n$, respectively, and each restriction $g|_{U_j}$ is injective and has a holomorphic inverse $g_j^{-1}:V \to U_j$, $j = 1, \cdots, n$. Therefore

$$\sum_{j=1}^{n} \iint_{U_j} A(\zeta)|g'(\zeta)|^2\, d\xi d\eta = \sum_{j=1}^{n} \iint_V A \circ g_j^{-1}(w)\, dx^1 dx^2$$

$$= \iint_V \frac{384a^3(1-|w|^2)|w|^4}{(1+|w|^2)^7} \cdot \frac{F(w)}{w^2}\, dx^1 dx^2 .$$

Consequently

$$\iint_{D_1} A(\zeta)|g'(\zeta)|^2\, d\xi d\eta = \iint_{D_2} \frac{384a^3(1-|w|^2)|w|^4}{(1+|w|^2)^7} \cdot \frac{F(w)}{w^2}\, dx^1 dx^2 . \qquad (24)$$

Now we assert that each integrand of the both sides of (24) is bounded. In fact, because of the regularity and the extensibility of x, $1/f$ and g' are both bounded on $\bar{D}$ (cf. §2). Moreover, not only $D - D_1$, but also $D_0 - D_2$ is of measure zero. Therefore (24) implies (23).

Let us introduce the polar coordinates (r, θ) in w-plane : $w = x^1 + \sqrt{-1}\, x^2 = re^{\sqrt{-1}\,\theta}$. By using (23) and the residue theorem, we obtain

$$I^{(3)}(u) = \mathrm{Re} \int_{r=0}^{1} \frac{384a^3(1-r^2)r^5}{(1+r^2)^7} \cdot \frac{1}{\sqrt{-1}} \left(\int_{|w|=r} \frac{F(w)}{w^3}\, dw \right) dr$$

$$= \mathrm{Re} \int_{r=0}^{1} \frac{384a^3(1-r^2)r^5}{(1+r^2)^7} \cdot \frac{1}{\sqrt{-1}} \cdot 2\pi\sqrt{-1} \cdot \operatorname*{Res}_{w=0} \left\{ \frac{F(w)}{w^3} \right\} dr$$

which implies (22). Q.E.D.

Therefore, by taking account of Remark 2 and Lemma 9, we obtain

Main theorem. Assume that the Gauss map $\mathfrak{G}$ of a regular minimal surface $x:\bar{D} \to \mathbb{R}^3$ is a mapping from $\bar{D}$ onto the lower hemisphere H^- of S and that $\mathfrak{G}$ maps ∂D onto ∂H^-. If

$$\mathrm{Re}\,\{F''(0)\} \neq 0,$$

then x is unstable, where F is the function defined by (21).

REFERENCES

[1] J.L.Barbosa and M.do Carmo, On the size of a stable minimal surface in $\mathbb{R}^3$, Amer. J. Math., 98 (1976), 515-528.

[2] M.Beeson, Some results on finiteness in Plateau's problem, Part I, Math. Z., 175 (1980), 103-123.

[3] R.C.Gunning, Lectures on Riemann surfaces, Princeton Math. Notes, Princeton Univ. Press, Princeton, 1966.

[4] J.C.C.Nitsche, Vorlesungen über Minimalflächen, Springer, Berlin-Heiderberg-New York, 1975.

[5] J.Peetre, A generalization of Courant's nodal domain theorem, Math. Scand., 5 (1957), 15-20.

COMPACT HOMOGENEOUS SUBMANIFOLDS
WITH PARALLEL MEAN CURVATURE

Yoshihisa Kitagawa

1. Introduction.

Let $\mathbb{R}^n$ be the n-dimensional Euclidean space. It seems to be interesting to consider the problem: What are the compact submanifolds in $\mathbb{R}^n$ with parallel mean curvature ? This problem has been studied by several authors under some suitable assumptions. In this note we consider the above problem under the assumption that the submanifolds are homogeneous, where a submanifold M in $\mathbb{R}^n$ is said to be homogeneous if M is an orbit of an isometry group of $\mathbb{R}^n$. The following examples are known.

Example 1. Let M be a compact homogeneous submanifold in $\mathbb{R}^n$. If $n - \dim M \leqq 2$ then M is a compact homogeneous submanifold in $\mathbb{R}^n$ with parallel mean curvature.

Example 2. Let M be a compact homogeneous minimal submanifold in a hypersphere of $\mathbb{R}^n$. Then M is a compact homogeneous submanifold in $\mathbb{R}^n$ with parallel mean curvature.

Example 3 (Kitagawa and Ohnita [2]). Let G/K be a Riemannian symmetric space with the origin $o = K$. Then K acts on the tangent space $T_o(G/K)$ as an isometry group through the isotropy representation. If M is an orbit of K then M is a compact homogeneous submanifold in $T_o(G/K)$ with parallel mean curvature.

Example 4. For $1 \leqq i \leqq k$, let $M_i \subset \mathbb{R}^{n_i}$ be a submanifold which is given in Examples 1 - 3. Then the product $M_1 \times \cdots \times M_k \subset \mathbb{R}^{n_1} \times \cdots \times \mathbb{R}^{n_k}$ is a compact homogeneous submanifold with parallel mean curvature.

It is natural to consider the possibility of other examples. Hence we ask the following question.

Question: Does there exist any m-dimensional compact homogeneous submanifold M in $\mathbb{R}^{m+3}$ with parallel mean curvature which is not given in Examples 2 - 4 ?

In the remaining sections we will try to answer this question.

2. Preliminaries.

Let G be a compact connected Lie subgroup of $SO(n)$. Then G acts naturally on $\mathbb{R}^n$ as an isometry group. For each point $x \in \mathbb{R}^n$ let $G(x)$ denote the G-orbit of x. We put as follows:

$$r(G) = n - \max\{\dim G(x) : x \in \mathbb{R}^n\}.$$

The number $r(G)$ is called the cohomogeneity of G. A point $x \in \mathbb{R}^n$ is said to be a regular point of G if $r(G) = n - \dim G(x)$. It is known that the set of all regular points of G is dence in $\mathbb{R}^n$.

Lemma 2.1. Let G_1 and G_2 be compact connected Lie subgroups of $SO(n)$. If $G_1 \subset G_2$ and $r(G_1) = r(G_2)$ then $G_1(x) = G_2(x)$ for all $x \in \mathbb{R}^n$.

Proof. (Step 1) Let $x \in \mathbb{R}^n$ be a regular point of G_1. Then we have $\dim G_1(x) = n - r(G_1) = n - r(G_2) \geqq \dim G_2(x)$. Since $G_1 \subset G_2$ we have $G_1(x) = G_2(x)$.
(Step 2) Assume that $G_1(x) \neq G_2(x)$ for some point $x \in \mathbb{R}^n$. Then there exists an element $g \in G_2$ such that $G_1(x) \cap G_1(gx) = \phi$. Let d be the distance between $G_1(x)$ and $G_1(gx)$. Since the set of all regular points of G_1 is dense in $\mathbb{R}^n$, there exists a regular point $y \in \mathbb{R}^n$ of G_1 such that the distance between x and y is smaller than $d/2$. Then it is easy to show that $G_1(y) \cap G_1(gy) = \phi$. On the other hand (Step 1) implies that $G_1(y) = G_2(y) = G_2(gy) \supset G_1(gy)$. This is a contradiction.

Lemma 2.2. Let $S^{n-1} = \{x \in \mathbb{R}^n : \langle x,x\rangle = 1\}$ and let G be a compact connected Lie subgroup of $SO(n)$ such that $r(G) = 2$. If $x \in S^{n-1}$ and $\dim G(x) \leqq n - 3$ then the orbit $G(x)$ is a minimal submanifold in S^{n-1}.

Proof. For the submanifold $G(x)$ in $\mathbb{R}^n$ we denote by H and $N(G(x))$ the mean curvature vector field and the normal bundle, respectively. Let G_x be the isotropy subgroup of x. Then the group G_x acts naturally on the normal space $N_x(G(x))$ at x. This action, denoted by ϕ_x, satisfies the following.

$$\phi_x(g)\vec{x} = \vec{x} \quad \text{for all} \quad g \in G_x, \tag{2.1}$$

where $\vec{x}$ is the normal vector at x defined by $\vec{x} = \frac{d}{dt}(x + tx)|_{t=0}$.

There exists a positive number ε such that $\exp: N^{\varepsilon}(G(x)) \longrightarrow \mathbb{R}^n$ is an into diffeomorphism, where $N^{\varepsilon}(G(x)) = \{v \in N(G(x)): |v| < \varepsilon\}$. We choose a regular point $y \in \mathbb{R}^n$ of G such that $y \in \exp(N^{\varepsilon}(G(x)))$ and let ξ be the element of $N^{\varepsilon}(G(x))$ such that $\exp(\xi) = y$. We may assume that $\xi \in N_x(G(x))$. Then the isotropy subgroup G_y satisfies the following.

$$G_y = \{g \in G_x: \phi_x(g)\xi = \xi\}. \tag{2.2}$$

Let $V = \{v \in N_x(G(x)): \langle v,\vec{x}\rangle = 0\}$ and let $\xi_0 = \xi - \langle\xi,\vec{x}\rangle\vec{x}$. Then (2.1) and (2.2) imply that $G_y = \{g \in G_x: \phi_x(g)\xi_0 = \xi_0\}$. Since $n - 2 = \dim G(y)$ and $\dim G(x) \leqq n - 3$, we have $\dim(G_x/G_y) = \dim V - 1$ and $\dim V \geqq 2$. Hence the orbit $G_x(\xi_0)$ is a hypersphere of V and an element of V which is fixed for all $\phi_x(g)$ must be zero. This implies that a normal vector at x which is fixed for all $\phi_x(g)$ must be proportional to $\vec{x}$. On the other hand the mean curvature vector $H(x)$ is fixed for all $\phi_x(g)$. Hence $H(x)$ is proportional to $\vec{x}$. This implies Lemma 2.2.

3. Statement of result.

For $m \geqq 1$, let M be a m-dimensional compact connected homogeneous submanifold in $\mathbb{R}^{m+3}$ with parallel mean curvature. Let G_M be the largest connected Lie subgroup of $ISO(\mathbb{R}^{m+3})$ leaving M invariant. Then M is an orbit of G_M and we may assume that G_M is a compact connected Lie subgroup of $SO(m+3)$. It is easy to see that the cohomogeneity of G_M is 2 or 3. If $r(G_M) = 2$ then by Lemma 2.2 the submanifold $M \subset R^{m+3}$ is given in Example 2.

Now we consider the case that $r(G_M) = 3$. Using Lemma 2.1, the results of Hsiang and Lawson [1] and Uchida [3] imply that the submanifold $M \subset R^{m+3}$ is congruent to an orbit of the following representations (G,ϕ).

(1) G is a circle group acting on $\mathbb{R}^4$.

(2) $\phi = \phi' + \theta$, (G,ϕ') is a compact linear group of cohomogeneity 2 and θ is a 1-dimensional trivial representation.

(3) $G = SO(k) \times G'$, $\phi = \rho_k + \phi'$ for $k \geqq 2$, and (G',ϕ') is a compact linear group of cohomogeneity 2.

(4) $G = SO(k)$ and $\phi = 2\rho_k$ for $k \geqq 3$.

(5) $G = Sp(k) \times Sp(1) \times Sp(1)$ and $\phi = \nu_k \otimes_{\mathbb{Q}} (\nu_2^* | Sp(1) \times Sp(1))$ for $k \geqq 2$.

(6) $G = SU(k) \times U(1) \times U(1)$ and $\phi = [\mu_k \otimes_{\mathbb{C}} (\mu_2^* | U(1) \times U(1))]_{\mathbb{R}}$ for $k \geqq 2$.

(7) $G = Spin(9)$ and $\phi = \Delta_9 + \rho_9$.

(8) $G = Sp(2) \times Sp(1)$, $\phi = (\nu_2 \otimes_{\mathbb{Q}} \nu_1^*) + \pi$, and $\pi : Sp(2) \longrightarrow SO(5)$ is a surjection.

(9) $G = U(2)$, $\phi = [\mu_2]_{\mathbb{R}} + \pi'$, and $\pi' : U(2) \longrightarrow SO(3)$ is a surjection.

(10) $G = SO(2) \times Spin(9)$ and $\phi = \rho_2 \otimes_{\mathbb{R}} \Delta_9$.

(11) (G,ϕ) is the isotropy representation of an irreducible symmetric space of rank 3.

It is easy to see the following.

(A) Let M be a codimension 3 orbit of the representation (1) with parallel mean curvature. Then M is a circle in $\mathbb{R}^4$.

(B) Let M be a codimension 3 orbit of the representations (2), (3). Then M is a submanifold which is given in Example 4.

(C) Let M be an orbit of the representation (11). Then M is a submanifold which is given in Example 3.

For the other representations we have the following result.

Theorem. Let M be a codimension 3 orbit of the representations (4)-(10) with parallel mean curvature. Then M is a submanifold which is given in Example 2.

4. Proof of Theorem.

In this section we prove the assertion of Theorem for the representation (5). In the remaining cases, the assertions of Theorem can be proved similarly.

Let $\mathbb{Q}$ be the algebra of quaternions and let $M(k,2;\mathbb{Q})$ be the set of all $k\times 2$ quaternionic matrices ($k \geqq 2$). For $X, Y \in M(k,2;\mathbb{Q})$ we put as follows:

$$\langle X,Y\rangle = \operatorname{trace}(X^*Y), \quad \operatorname{Re}\langle X,Y\rangle = \text{real part of } \langle X,Y\rangle,$$

where X^* is the transpose of the conjugate of X. Then $M(k,2;\mathbb{Q})$ is an $8k$-dimensional Euclidean vector space with the inner product $\operatorname{Re}\langle\,,\rangle$. Let $G = Sp(k) \times Sp(1) \times Sp(1)$ and let ϕ be a real representation of G on $M(k,2;\mathbb{Q})$ defined by the following.

$$\phi(A,q_1,q_2)X = AX\begin{bmatrix}\bar{q}_1 & 0\\ 0 & \bar{q}_2\end{bmatrix},$$

where $A \in Sp(k)$, $q_i \in Sp(1)$, $X \in M(k,2;\mathbb{Q})$. Then the inner product $\mathrm{Re}\langle\ ,\ \rangle$ is (G,ϕ)-invariant and $\phi = \nu_k \otimes_{\mathbb{Q}} (\nu_2^* | Sp(1) \times Sp(1))$. Let $\mathfrak{g} = \mathfrak{sp}(k) + \mathfrak{sp}(1) + \mathfrak{sp}(1)$ be the Lie algebra of G and let ϕ_* be the representation of $\mathfrak{g}$ induced by ϕ. We put as follows:

$$S^{8k-1} = \{X \in M(k,2;\mathbb{Q}) : \mathrm{Re}\langle X,X\rangle = 1\},$$

$$\Omega = \{{}^t\begin{bmatrix} a & 0\cdots 0\\ b & c\cdots 0\end{bmatrix} \in M(k,2;\mathbb{R}) : \begin{matrix} a^2 + b^2 + c^2 = 1\\ a \geqq 0,\ b \geqq 0,\ c \geqq 0\end{matrix}\},$$

$$\mathring{\Omega} = \{{}^t\begin{bmatrix} a & 0\cdots 0\\ b & c\cdots 0\end{bmatrix} \in \Omega : a > 0,\ b > 0,\ c > 0\}.$$

Then straightforward computations imply the following.

Lemma 4.1. If $X \in S^{8k-1}$ then the orbit $G(X)$ meets Ω.

Lemma 4.2. Let $X \in \Omega$. Then $\dim G(X) = 8k - 3$ if and only if $X \in \mathring{\Omega}$.

Let M be an $(8k - 3)$-dimensional orbit of the representation (G,ϕ) with parallel mean curvature. In order to prove that M is a minimal submanifold in a hypersphere of $M(k,2;\mathbb{Q})$, we may assume that $M \subset S^{8k-1}$. Hence by Lemmas 4.1 and 4.2 there exists an element $Z \in \mathring{\Omega}$ such that $M = G(Z)$. The tangent space T_ZM is naturally identified with a linear subspace of $M(k,2;\mathbb{Q})$. Then we have the following.

(4.1) $T_ZM = \{X = (x_{ij}) : x_{ij} = -\bar{x}_{ij} \text{ for } 1 \leqq i, j \leqq 2\} + \mathrm{span}_{\mathbb{R}} Z'$,

where $Z = {}^t\begin{bmatrix} a & 0\cdots 0\\ b & c\cdots 0\end{bmatrix}$ and $Z' = {}^t\begin{bmatrix} 0 & a\cdots 0\\ -c & b\cdots 0\end{bmatrix}$. Let N_ZM be the orthogonal complement of T_ZM in $M(k,2;\mathbb{Q})$. We put as follows:

$$E_0 = Z,\quad E_1 = {}^t\begin{bmatrix} c & b\cdots 0\\ 0 & -a\cdots 0\end{bmatrix},\quad E_2 = {}^t\begin{bmatrix} -b & c\cdots 0\\ a & 0\cdots 0\end{bmatrix},$$

$$\eta = \left(\begin{bmatrix} \begin{matrix}0 & -1\\ 1 & 0\end{matrix} & 0\\ 0 & 0\end{bmatrix},\ 0,\ 0\right) \in \mathfrak{sp}(k) + \mathfrak{sp}(1) + \mathfrak{sp}(1).$$

Then simple computations and (4.1) imply the following.

Lemma 4.3. $\{E_0, E_1, E_2\}$ is an orthonormal basis of N_ZM such that $\phi_*(\eta)E_0 \in T_ZM$, $\phi_*(\eta)E_1 = E_2$ and $\phi_*(\eta)E_2 = -E_1$.

For the submanifold M in $M(k,2;\mathbb{Q})$ we denote by H and D the mean curvature vector field and the normal connection, respectively. Define $X \in T_Z M$ by $X = \phi_*(\eta)Z$. Then it is easy to show that $D_X H = N_Z M$-component of $\phi_*(\eta)H(Z)$. Hence by Lemma 4.3 we have the following.

$$D_X H = \mathrm{Re}\langle H(Z),E_1\rangle E_2 - \mathrm{Re}\langle H(Z),E_2\rangle E_1.$$

Therefore the assumption that $DH = 0$ implies that $\mathrm{Re}\langle H(Z),E_i\rangle = 0$ for $1 \leqq i \leqq 2$. This shows that $H(Z)$ is proportional to Z. Hence M is a minimal submanifold in S^{8k-1}. This completes the proof of Theorem.

References

[1] W.Y. Hsiang and H.B. Lawson,Jr., Minimal submanifolds of low cohomogeneity, J. Diff. Geometry 5(1971), 1-38.

[2] Y. Kitagawa and Y. Ohnita, On the mean curvature of R-spaces, Math. Ann. 262(1983), 239-243.

[3] F. Uchida, An orthogonal transformation group of (8k - 1)-sphere, J. Diff. Geometry 15(1980), 569-574.

Department of Mathematics
Utsunomiya University
Mine-machi, Utsunomiya, 321 Japan

SUR LES ENSEMBLES NODAUX

Kinji Watanabe

Univertsité Educationale d'Hyogo

Yashiro, Hyogo, 673-14, Japon

1. Théorèmes d'unicité forte.

On sait bien le résultat d'Aronszajn et Cordes concernant l'unicité forte à partir d'un point de l'intérieur pour les équations elliptiques du second ordre (pour les équations à coefficients complexes, voir Alinhac-Baouendi [1]). Nous énonçons ici un théorème d'unicité forte à partir d'un point de la frontière. Soit L un opérateur différentiel défini dans un domaine borné Ω de R^n, à bord Γ de classe C^∞, de la forme suivante :

$$(1.1)\qquad L[u] = \sum_{i,j=1}^{n} \frac{\partial}{\partial x_i}\left(a_{i,j}(x)\frac{\partial u}{\partial x_j} \right) + \sum_{j=1}^{n} a_j(x)\frac{\partial u}{\partial x_j} + a(x)u .$$

Ici tous les coefficients sont de classe $C^\infty(\overline{\Omega})$, $a_{i,j} = a_{j,i} = \overline{a}_{i,j}$ et la forme quadratique $(a_{i,j}(x))$ est strictement positif pour tout x dans $\overline{\Omega}$.

Théorème 1.

Soit u une solution de l'équation $L[u] = 0$ dans Ω vérifiant la condition de Dirichlet : $u = 0$ sur Γ ou bien la condition de Neumann :

$$\sum_{i,j=1}^{n} a_{i,j}(x)\, \nu_i(x)\, \frac{\partial u}{\partial x_j} = 0 \quad \text{sur } \Gamma$$

où $(\nu_1(x),\dots,\nu_n(x))$ est le vecteur normal unité en x de Γ. Si u s'annule à l'ordre infini en un point x_o de Γ, elle est identiquement nulle dans Ω.

Remarquons que ce Théorème est aussi valide au cas où u vérifie la condition frontière de la forme suivante :

$$\sum_{j=1}^{n} b_j(x) \frac{\partial u}{\partial x_j} + b(x)u = 0 \quad \text{sur } \Gamma$$

au lieu de la condition de Neumann. Ici les coefficients sont de classe C^∞, $b_j = \overline{b}_j$ et $b_j(x_o) = \sum_{j=1}^{n} a_{i,j}(x_o)\nu_i(x_o)$ pour $j = 1,\ldots,n$.

Il est notable qu'il existe des fonctions harmoniques s'annulant à l'ordre infini en un point de la limite. Un exemple de telle fonction est donné par la partie réelle de $\exp(-1/z^{1/2})$ où $z = x_1 + \sqrt{-1}\, x_2$.

En considérant des extensions convenables de u près de x_o, le Théorème 1 est une conséquence d'un théorème d'unicité forte à partir d'une ligne dans l'intérieur pour des équations paraboliques associées aux équations elliptiques à coefficients non lipschitziens. Il s'agit d'une solution v de l'inégalité différentielle dans un voisinage $\omega \times]-T,T[$ de l'origine de R^{n+1} de la forme suivante :

$$(1.2) \quad \left|\frac{\partial v}{\partial t} + \sum_{i,j=1}^{n} A_{i,j}(x,t)\frac{\partial^2 v}{\partial x_i \partial x_j}\right| \leq C\left\{ |x|^{-3/2}|v| + |x|^{-1/2}\sum_{j=1}^{n}\left|\frac{\partial v}{\partial x_j}\right| \right\}.$$

Nous faisons les hypothèses suivantes :

(1.3) Tous les coefficients sont mésurables et leurs premières dérivées partielles existent presque partout et sont bornées. De plus il existe une constante $C_1 > 0$ telle que pour $i,j = 1,\ldots,n$

$$| A_{i,j}(x,t) - \delta_{i,j} | \leq C_1|x| , \quad \text{p.p. dans } \omega \times]-T,T[$$

où $\delta_{i,j}$ est le delta de Kronecker.

(1.4) La solution v et ses dérivées partielles $\partial v/\partial t$, $\partial v/\partial x_j$, $\partial^2 v/\partial x_i \partial x_j$, ($i,j = 1,\ldots,n$) appartiennent à $L^2(\omega \times]-T,T[)$.

<u>Théorème 2</u>.

Soit v une solution de (1.2) vérifiant (1.3) et (1.4).

Si v s'annule à l'ordre infini sur $\{0\}\times]-T,T[$ au sens suivant :

$$\sum_{|\alpha|\leq 2} \int_{-T}^{T} \int_{\omega} \left| \left(\frac{\partial}{\partial x}\right)^{\alpha} v(x,t) \right|^2 |x|^{-\tau} \, dx\, dt < \infty \quad \text{pour tout } \tau > 0,$$

elle est identiquement nulle près de l'origine.

La preuve est basé sur l'estimation du type Carleman avec fonctions de poids $|x|^{-2\tau}$. Après un changement de variables singulrière $(x,t) \to (y,t)$ défini par $y = \eta(\xi(t)x)\ \xi(t)x$, où $\eta(r) = 1 - \sigma r$, $\xi(t) = \exp(-1/(t^2 - T_1^2))$, $0 < T_1 < T$, $\sigma > 0$ est un grand paramètre, un opérateur apparu dans (1.2) $\xi(t)^2\{\partial/\partial t + \sum_{i,j=1}^{n} A_{i,j}(x,t)\, \partial^2/\partial x_i \partial x_j\}$ se transforme à un opérateur, dont " la partie princiale " est de la forme suivante :

$$P_\sigma[\phi] = \xi(t)^2 \frac{\partial \phi}{\partial t} + \sum_{i,j=1}^{n} A_{i,j,\sigma}(y,t) \frac{\partial^2 \phi}{\partial y_i \partial y_j}$$

et nous énonçons l'inégalitésuivante,qui implique l'unicité exigée.

Proposition 3.

Pour assez grand σ, il existe des constantes ε_o, τ_o, $C_o > 0$ telles que pour tout $0 < \varepsilon < \varepsilon_o$, tout $\tau > \tau_o$ et tout ϕ dans $C^\infty(R^{n+1})$ vérifiant $\phi = 0$ en dehors de $\{(y,t)\ ; |y| = r \leq \varepsilon,\ |t| \leq T_1\}$ et s'annulant à l'ordre infini sur $\{0\}\times R^1$:

$$C_o \| r^{-\tau} P_\sigma[\phi] \|^2 \geq \tau \sum_{j=1}^{n} \| r^{-\tau-(1/2)} \frac{\partial \phi}{\partial y_j} \|^2 + \tau^3 \| r^{-\tau-(3/2)} \phi \|^2.$$

Ici $\|\ \|$ est la norme usuelle dans $L^2(R^{n+1})$.

2. Ensembles nodaux des fonctions propres.

Dans ce paragraphe nous supposons que tous les coefficients de (1.1) soient à valeurs réelles et considérons une solution u à valeurs réelles de l'équation $L[u] = 0$ dans Ω et son ensemble nodal $V(u) = \{ x \in \Omega\ ;\ u(x) = 0 \}$. Soit Ω' un domaine nodal de u, i.e.

un composant de $\Omega \setminus V(u)$ et posons $u'(x) = u(x)$ dans Ω', $= 0$ dans $\Omega \setminus \Omega'$. Pour complèter la démonstration du théorème 2.2 de Cheng [4] nous donnons une proposition suivante,qui permet de retrouver le théorème de Courant concernant le nombre des domaines nodaux des fonctions propres en dimension générale sans aucune hypothèse sur régularité des domaines nodaux (voir Bérard [2] , Cheng [4] et Courant-Hilbert[6]).

Proposition 4.

$\{ x \in V(u) \; ; \; \operatorname{grad} u(x) = 0 \}$ est un sous-ensemble d'une réunion du nombre nombrable au plus des C^∞ variétés de dimension $n - 2$ et pour $j = 1,\dots,n$, $\partial u'/\partial x_j = (\partial u/\partial x_j)'$ au sens de distribution dans Ω .

Preuve. Pour $k > 1$ posons

$$V_k = \{x \in V(u) ; \sum_{|\alpha|<k} |(\frac{\partial}{\partial x})^\alpha u(x)| = 0, \sum_{|\alpha|=k} |(\frac{\partial}{\partial x})^\alpha u(x)| > 0\}$$

et prenons un point x_o dans V_k et des coordonnées normales centrées en x_o , i.e. $a_{i,j}(x_o) = \delta_{i,j}$. Puisque le polynôme

$$p(x) = \sum_{|\alpha|=k} (1/\alpha!)(x - x_o)^\alpha (\frac{\partial}{\partial x})^\alpha u(x_o) \tag{2.1}$$

est harmonique, V_k est , près de x_o , un sous-ensemble d'une C^∞ variété de dimension $n - 2$ défini par $\partial/\partial x_i (\partial/\partial x)^\beta u(x) = \partial/\partial x_j (\partial/\partial x)^\beta u(x) = 0$ pour des certaines indices $i \neq j$, β avec $|\beta| = k - 2$ ($\operatorname{grad} \partial/\partial x_i (\partial/\partial x)^\beta u(x_o)$ et $\operatorname{grad} \partial/\partial x_j (\partial/\partial x)^\beta u(x_o)$ sont linéairement indépendants parceque le rang de l'hessienne de $(\partial/\partial x)^\beta p$ est égal ou supérieur à 2 au cas de $(\partial/\partial x)^\beta p \not\equiv 0$). Il'en résulte que le support de $\partial u'/\partial x_j - (\partial u/\partial x_j)'$ ne contient aucun point de V_k.

Nous considérons ensuite l'ensemble nodal près d'un point de Γ au cas de $n = 2$. La proposition suivante est une conséquence du Théorème 1 et des résultats de Cheng [4], qui a cherché en détail $V(u)$ près d'un point de l'intérieur.

Proposition 5.

Soit u une solution non triviale de l'équation $L[u] = 0$ dans Ω ,vérifiant une des conditions de Dirichlet et de Neumann et s'annulant en un point x_o de Γ . Il existe alors un système équiangulaire en x_o par rapport aux coordonnées convenables, $\{\gamma_j \; ; \; j = 1,\ldots,k\}$ (k = l'ordre d'annulation en x_o) des C^∞ curves tel que si u satisfait à la condition de Dirichlet, on ait

$$V(u) = \bigcup_{j=1}^{k-1} (\gamma_j \cap \Omega) , \quad \gamma_k = \Gamma \quad \text{près de } x_o$$

et que si u satisfait à la condition de Neumann, on ait

$$V(u) = \bigcup_{j=1}^{k} (\gamma_j \cap \Omega) \quad \text{près de } x_o$$

et Γ est divise en deux partie égales l'angle entre γ_1 et γ_k , qui est égale à π/k.

Preuve. En vertu du Théorème 1 nous avons que $k < \infty$ et le Lemme 2.4 de [4] résulte donc qu'une C^∞ extension de u est C^1 équivalente au polynôme p défini par (2.1). Il est alors aisé de décrire $V(u)$ par les lignes nodales de p.

3. Multiplicités des valeurs propres.

Dans ce paragraphe nous supposons que $n = 2$ et que $a_1 = \ldots = a_n = a = 0$ et considérons les multiplictés des valeurs propres du problème de Dirichlet (resp. Neumann) :

$$(3.1) \quad L[u] \equiv \sum_{i,j=1}^{2} \frac{\partial}{\partial x_i} \left(a_{i,j}(x) \frac{\partial u}{\partial x_j} \right) = - \lambda u \quad \text{dans } \Omega$$

$$u = 0 \text{ sur } \Gamma \quad \left(\text{ resp. } \sum_{i,j=1}^{2} a_{i,j}(x)\, \nu_i(x) \frac{\partial u}{\partial x_j} = 0 \text{ sur } \Gamma\right).$$

Nous écrivons

$$0 < \lambda_1 < \lambda_2 \leqq \lambda_3 \leqq \ldots, \quad (\text{resp. } 0 = \mu_0 < \mu_1 \leqq \mu_2 \leqq \ldots\,,)$$

la suite des valeurs propres (écrités avec multiplicités) pour ce problème (3.1). Nous donnons un résultat pareil aux Bessons [3] et Cheng [4] .

<u>Proposition 6</u>.

Soit $m_j(D)$ (resp. $m_j(N)$) la multiplicité de λ_j (resp. μ_j). On a alors pour $j \geqq 1$

$$m_j(D) \leqq 2j - 1, \quad m_j(N) \leqq 2j + 1.$$

<u>Lemme 7</u>.

Soit u une fonction propre non triviale de (3.1) associée à une valeur propre λ_j (resp. μ_j) . Elle s'annule alors à l'ordre $\leqq j - 1$ (resp. j) en chaque point de Ω et à l'ordre k avec $[k/2] \leqq j - 1$ (resp. $[(k + 1)/2] \leqq j$) en chaque point de Γ.

<u>Preuve du Lemme 7</u>. En utilisant le fait que R^2 est simplement connexe ,le résultat de Cheng [4] et la proposition 5 donnent des minorations du nombre des domaines nodaux de u en fonction de l'ordre d'annulation. D'autre part ce nombre est inférieur à j (resp. $j+1$) du théorème de Courant. Ceci permet de terminer la preuve.

<u>Preuve de la proposition 6.</u> En utilisant une méthode de Cheng [4] et Besson [3] , cette proposition est une conséquence immédiate du Lemme 7 et du théorème de Courant.

Nous finalement remarquons que pour les variétés compactes riemanniennes de dimension quelconque , M , on a pour $j \geq 1$

$$m_j(M) \leq C_1(M)j + C_2(M)$$

où $m_j(M)$ est la multiplicité de la j-ième valeur propre positif $\lambda_j(M)$ du laplacien de M et $C_k(M)$, $k = 1,2,$ sont des constantes dépendantes de M. Parce qu' une modification des calcules dans Cheng et Li [5] donne l'inégalité suivanté :

$$m_j(M) \leq C_3(M) \lambda_j(M)^{n/2}$$

et nous savons du résultat de Li et Yau [7] que

$$\lambda_j(M) \leq C_4(M) j^{2/n} + C_5(M).$$

BIBLIOGRAPHIE

[1] Alinhac,S. et Baouendi,M.S. Uniqueness for characteristic Cauchy problem and strong unique continuation for higher order partial differential inequalities, Amer. J. Math. , vol. 102, (1980) , 179-217.

[2] Bérard,P. Inégalités isopermétriques et applications, domaines nodaux des fonctions propres, Séminaire Goulaouic-Meyer-Schwartz, 1981-1982, Exposé XI.

[3] Besson,G. Sur la multiplicité de la première valeur propre des surfaces riemanniennes, Ann. Inst. Fourier, Grenoble, 30, (1980), 109-128.

[4] Cheng,S.Y. Eigenfunctions and nodal sets, Comment. Math. Helvetici, 51, (1976), 43-55.

[5] Cheng,S.Y. et Li,P. Heat kernel estimates and lower bound of eigenvalues, Comment. Math. Helvetici, 56, (1981),327-338.

[6] Courant,R. et Hilbert,D. Methods of mathematical physics, vol. 1, New york, Interscience, 1953.

[7] Li,P. et Yau,S.T. Estimates of eigenvalues of a compact riemannian manifold, Proceedings of Symposia in pure mathematics, vol. 36, (1980), 205-239.

ON SOME STABLE MINIMAL CONES IN $\mathbb{R}^7$

Katsuya Mashimo
Institute of Mathematics
University of Tsukuba, Ibaraki 305, Japan

The 7-dimensional Euclidean space $\mathbb{R}^7$ can be considered as the space of all pure imaginary numbers of the Cayley algebra. Using the multiplicative structure of the Cayley algebra Harvey and Lawson defined two classes of submanifolds in $\mathbb{R}^7$, namely the class of associative submanifolds and the class of coassociative submanifolds. From a general Theorem, they are stable minimal submanifolds. In [4], Harvey and Lawson gave an example of coassociative submanifold. Their example is a cone over an orbit of a closed 3-dimensional subgroup of G_2. In this note we study coassociative submanifolds which are cones over orbits of 3-dimensional closed subgroups of G_2.

1. Preliminalies

Let M be an n-dimensional Riemannian manifold and let $G_p(M)$ be the bundle of oriented p-planes to M. Then we can consider $G_p(M)$ as a subset of the p-th exterior power $\Lambda^p(M)$ of the tangent bundle of M in a natural manner. Then any exterior p-form on M can be viewed as a function on $G_p(M)$. The <u>comass</u> of an exterior p-form ϕ is defined by

$$\|\phi\|^* = \sup_{\xi \in G_p(M)} \phi(\xi).$$

Assuming $\|\phi\|^*=1$, we put

$$G(\phi) = \{ \xi \in G_p(M) \mid \phi(\xi) = 1 \}.$$

A p-dimensional, oriented C^1-submanifold S of M is called a <u>ϕ-manifold</u> if the oriented p-plane $T_x(S)$ is contained in $G(\phi)$ for all $x \in S$.

<u>Theorem 1.1</u>(Harvey-Lawson, [4]). <u>Let</u> ϕ <u>be a closed p-form with</u>

$\|\phi\|^* = 1$ and S be a ϕ-manifold. Then for any compactly supported variation S_t of S

$$\mathrm{Vol}(S) \leqq \mathrm{Vol}(S_t)$$

holds, i.e., S is a stable minimal submanifold.

Let M be a submanifold in the n-dimensional unit sphere S^n in $\mathbb{R}^{n+1}$ centered at the origin. Then the cone CM over M is

$$CM = \{ tx \mid x \in M, \ t \geqq 0 \} .$$

2. Special geometries related to the Cayley algebra

Let $\mathbb{H}$ be the skew field of all quaternions. Then the Cayley algebra $\mathbb{C}a$ over $\mathbb{R}$ is $\mathbb{C}a = \mathbb{H} + \mathbb{H}$ with the following multiplication $\cdot$,

$$(q,r)\cdot(s,t) = (qs-\bar{t}r, \ tq+r\bar{s}), \quad q,r,s,t \in \mathbb{H},$$

where the symbol "$-$" means the conjugation in $\mathbb{H}$. We define a conjugation in $\mathbb{C}a$ by $\overline{(q,r)} = (\bar{q}, -r)$, $q,r \in \mathbb{H}$, and an inner product $\langle , \rangle$ by

$$\langle x, y\rangle = (x\bar{y} + y\bar{x})/2, \quad x, y \in \mathbb{C}a.$$

Let $1,i,j,k$ be the usual basis of $\mathbb{H}$. Then $e_0=(1,0)$, $e_1=(i,0)$, $e_2=(j,0)$, $e_3=(k,0)$, $e_4=(0,1)$, $e_5=(0,i)$, $e_6=(0,j)$, $e_7=(0,k)$ are orthonormal basis of $\mathbb{C}a$. We put

$$\mathbb{C}a_0 = \{ x \in \mathbb{C}a \mid x + \bar{x} = 0 \} = \textstyle\sum_{j=1}^{7} \mathbb{R}e_j .$$

We have the following multiplication table .

$e_i \cdot e_j =$

i/j	1	2	3	4	5	6	7
1	$-e_0$	e_3	$-e_2$	e_5	$-e_4$	$-e_7$	e_6
2	$-e_3$	$-e_0$	e_1	e_6	e_7	$-e_4$	$-e_5$
3	e_2	$-e_1$	$-e_0$	e_7	$-e_6$	e_5	$-e_4$
4	$-e_5$	$-e_6$	$-e_7$	$-e_0$	e_1	e_2	e_3
5	e_4	$-e_7$	e_6	$-e_1$	$-e_0$	$-e_3$	e_2
6	e_7	e_4	$-e_5$	$-e_2$	e_3	$-e_0$	$-e_1$
7	$-e_6$	e_5	e_4	$-e_3$	$-e_2$	e_1	$-e_0$

The Cayley algebra $\mathbb{C}a$ is not commutative nor associative. But we have the following

Lemma 2.1. (i) If $x, y \in \mathbb{C}a_0$, $x \cdot y = -y \cdot x$.
(ii) For any $x, y, z \in \mathbb{C}a$,

$$\bar{x} \cdot (x \cdot y) = (\bar{x} \cdot x) \cdot y,$$

$$\langle x \cdot y, x \cdot z \rangle = \langle x, x \rangle \langle y, z \rangle .$$

(iii) Let $x, y, z \in \mathbb{C}a$ be mutually orthogonal unit vectors, then

$$x \cdot (y \cdot z) = y \cdot (z \cdot x) = z \cdot (x \cdot y).$$

For the proof, we refer to [3].

It is well-known that the group of all automorphisms of $\mathbb{C}a$ is the compact connected simple Lie group of type $\mathfrak{g}_2$ [3]. So we denote it by G_2. Then G_2 leaves invariant the vector e_0 and the subspace $\mathbb{C}a_0$. Furthermore G_2 leaves invariant the inner product $\langle \ , \ \rangle$. So we may regard G_2 as a subgroup of $SO(7) = SO(\mathbb{C}a_0)$. Let G_{ij}, $1 \leqq i \neq j \leqq 7$, be the skew symmetric transformation on $\mathbb{C}a_0$ defined by

$$G_{ij}(e_k) = \begin{cases} e_i, & \text{if } k = j, \\ -e_j, & \text{if } k = i, \\ 0, & \text{if otherwise.} \end{cases}$$

Then the Lie algebra $\mathfrak{g}_2$ of G_2 is spanned by the following vectors in the Lie algebra $\mathfrak{so}(7)$ of $SO(7)$.

$$\begin{aligned} &aG_{23} + bG_{45} + cG_{76}, \\ &aG_{31} + bG_{46} + cG_{57}, \\ &aG_{12} + bG_{47} + cG_{65}, \\ &aG_{51} + bG_{73} + cG_{62}, \\ &aG_{14} + bG_{72} + cG_{36}, \\ &aG_{17} + bG_{24} + cG_{53}, \\ &aG_{61} + bG_{34} + cG_{25}, \quad a + b + c = 0,\ a, b, c \in \mathbb{R}. \end{aligned}$$

Let ϕ be a tri-linear map on $\mathbb{C}a_0$ defined by

$$\phi(x,y,z) = \langle x, y \cdot z \rangle, \quad x, y, z \in \mathbb{C}a_0.$$

Then we have the following

Proposition 2.2 (Harvey-Lawson, [4]). ϕ is a closed 3-form on $\mathbb{C}a_0$ with $\|\phi\|^* = 1$.

We fix an orientation on $\mathbb{C}a_0$ such that $e_1, e_2, \cdots, e_7$ is an oriented basis. Let $*$ be the Hodge star operator.

Proposition 2.3 (Harvey-Lawson, [4]). $*\phi$ is a closed 4-form on $\mathbb{C}a_0$ with $\|*\phi\|^* = 1$.

A ϕ-manifold is called an associative submanifold and a $*\phi$-manifold is called a coassociative submanifold.

Let S^6 be the unit sphere in $\mathbb{C}a_0$ centered at the origin. Then S^6 has an almost complex structure J defined by

$$J_p(X) = p\cdot X, \quad X \in T_p(S^6).$$

From the definition of the almost complex structure, G_2 preserves it. A submanifold M of S^6 is called a holomorphic submanifold (resp. totally real submanifold) if $J(T_p(M)) = T_p(M)$ (resp. $J(T_p(M))$ is contained in the normal space $N_p(M)$) for any $p \in M$.

Theorem 2.4. Let M be a 2-dimensional submanifold of S^6. Then M is a holomorphic submanifold if and only if the truncated cone $CM-\{o\}$ is an associative submanifold.

The proof of the above Theorem is easy.

Theorem 2.5. Let M be a 3-dimensional submanifold of S^6. Then M is a totally real submanifold if and only if the truncated cone $CM-\{o\}$ is a coassociative submanifold.

Proof. ($\Longleftarrow$). Let p be a point of M, and put $p = u_4$. Let u_5, u_6, u_7 be an orthonormal basis of $T_p(M)$ and u_1, u_2, u_3 be an orthonormal basis of the normal space $N_p(M)$ of M at p in S^6 such that $u_1, u_2, \cdots, u_7$ is oriented. Since $*\phi(u_4\wedge u_5\wedge u_6\wedge u_7) = \phi(u_1\wedge u_2\wedge u_3) = \pm 1$, we get $u_1\cdot u_2 = \pm u_3$. By Lemma 2.1, we get

$$\langle J(u_1), u_1\rangle = \langle u_4\cdot u_1, u_1\rangle = -\langle u_1\cdot u_4, u_1\rangle = -\langle u_4, \bar{u}_1\cdot u_1\rangle = 0,$$

$$<J(u_1),\ u_2> = <u_4\cdot u_1,\ u_2> = -<u_1\cdot u_4,\ u_2> = -<u_4,\ \bar{u}_1\cdot u_2> = 0,$$

$$<J(u_1),\ u_3> = <u_4\cdot u_1,\ u_3> = -<u_1\cdot u_4,\ u_3> = -<u_4,\ \bar{u}_1\cdot u_3> = 0,$$

i.e., $J(u_1)$ is contained in $T_p(M)$. Similary we see that $J(u_2)$ and $J(u_3)$ are contained in $T_p(M)$. Since J is non-singular and $\dim N_p(M) = \dim T_p(M)$, we get $J(T_p(M)) = N_p(M)$.
($\Longrightarrow$). Let u_5, u_6, u_7 be an orthonormal basis of $T_p(M)$. By a simple calculation, $(\nabla_{u_5} J)(u_6)$ is equal to the tangential part of $u_5\cdot u_6$ to S^6, where ∇ is the covariant derivative of S^6. Since M is a totally real submanifold, $u_5\cdot u_6$ is normal to p. In [2], Ejiri proved that $(\nabla_{u_5} J)(u_6)$ is normal to M. So $u_5\cdot u_6$ is normal to M. By Lemma 2.1. (ii), $u_5\cdot u_6$ is normal to $p\cdot u_5$ and $p\cdot u_6$. Since $N_p(M)$ is of dimension 3 and $u_5\cdot u_6$, $p\cdot u_7$ are unit vectors, we see $u_5\cdot u_6 = \pm p\cdot u_7$. Similarly $u_6\cdot u_7 = \pm p\cdot u_5$ and $u_7\cdot u_5 = \pm p\cdot u_6$. Then it is easily seen that $p\cdot u_5$, $p\cdot u_6$, $p\cdot u_7$ is an orthonormal basis of $T_p(M)$. So by Lemma 2.1,

$$\begin{aligned} *\phi(p\wedge u_5\wedge u_6\wedge u_7) &= \pm\phi(p\cdot u_5\wedge p\cdot u_6\wedge p\cdot u_7) = \pm<p\cdot u_5,\ (p.u_6)\cdot(p.u_7)> \\ &= \pm 1. \end{aligned}$$

So $CM-\{o\}$ is a coassociative submanifold. Q.E.D.

3. 3-dimensional closed subgroups of G_2

Let g be a compact simple Lie algebra and a be a maximal abelian subalgebra of g. Let ℓ be a complex simple 3-dimensional subalgebra of $g^{\mathbb{C}}$. Then there exists a basis H, X_+, X_- of ℓ such that

$$[H,\ X_+] = 2X_+, \quad [H,\ X_-] = -2X_-, \quad [X_+,\ X_-] = H.$$

We may assume that H is contained in $a^{\mathbb{C}}$, in fact in $\sqrt{-1}a$ [1,p.166]. So $\alpha(H)$ is a real number for every root α of $g^{\mathbb{C}}$ with respect to $a^{\mathbb{C}}$. Furthermore $\alpha(H) = 0$, 1 or 2 if α is a simple root. The weighted Dynkin diagram with weight $\alpha(H)$ added to each vertex α of the Dynkin diagram is called the <u>charasteristic diagram</u> of ℓ. There exists a one to one correspondence between charasteristic diagrams and complex simple 3-dimensional subalgebras up to inner automorphism of $g^{\mathbb{C}}$.

Mal'cev [5] classified the complex simple 3-dimensional subalgebra of $g_2^{\mathbb{C}}$. From his classification $g_2^{\mathbb{C}}$ has 4-types of 3-dimensional simple subalgebras. The charasteristic diagram of such subalgebra is one of the following.

I. $\overset{1}{\circ}\Rrightarrow\overset{0}{\circ}$

II. $\overset{0}{\circ}\Rrightarrow\overset{1}{\circ}$

III. $\overset{2}{\circ}\Rrightarrow\overset{0}{\circ}$

IV. $\overset{2}{\circ}\Rrightarrow\overset{2}{\circ}$

Let ℓ be a 3-dimensional simple subalgebra of g_2. Then the complexification $\ell^{\mathbb{C}}$ of ℓ in $g_2^{\mathbb{C}}$ corresponds to one of the above 4. If $\ell^{\mathbb{C}}$ corresponds to I, II or IV, then ℓ is unique up to inner automorphism of g_2. But if $\ell^{\mathbb{C}}$ correspond to III then there exists infinitely many subalgebras which are not conjugate in g_2. Now we give the basis of ℓ with $[X_1, X_2] = 2X_3$, $[X_2, X_3] = 2X_1$ and $[X_3, X_1] = 2X_2$.

<u>Case 1</u>. If $\ell^{\mathbb{C}}$ corresponds to $\overset{1}{\circ}\Rrightarrow\overset{0}{\circ}$, then

$$
\begin{aligned}
&X_1 = -G_{45} + G_{76},\\
(3.1)\qquad &X_2 = -G_{46} + G_{57},\\
&X_3 = -G_{47} + G_{65}.
\end{aligned}
$$

<u>Case 2</u>. If $\ell^{\mathbb{C}}$ corresponds to $\overset{0}{\circ}\Rrightarrow\overset{1}{\circ}$, then

$$
\begin{aligned}
&X_1 = -2G_{23} + G_{45} + G_{76},\\
(3.2)\qquad &X_2 = -2G_{31} + G_{46} + G_{57},\\
&X_3 = -2G_{12} + G_{47} + G_{65}.
\end{aligned}
$$

<u>Case 3.</u> If $\ell^{\mathbb{C}}$ corresponds to $\overset{2}{\circ}\Rrightarrow\overset{0}{\circ}$, then

$$
\begin{aligned}
&X_1 = -2G_{21} - 2G_{65}\\
(3.3)\qquad &X_2 = -2\cos\theta\,(G_{32} + G_{76}) - 2\sin\theta\,(G_{72} + G_{63}),\\
&X_3 = -2\cos\theta\,(G_{31} + G_{75}) - 2\sin\theta\,(G_{53} + G_{71}),
\end{aligned}
$$

where θ is a constant.

Case 4. If $\ell^{\mathbb{C}}$ corresponds to $\overset{2}{\circ}\Rrightarrow\overset{2}{\circ}$, then

$$X_1 = 4G_{32} + 2G_{54} - 6G_{76},$$

(3.4) $$X_2 = \sqrt{6}(G_{37} + G_{26} - 2G_{15}) + \sqrt{10}(G_{42} - G_{35}),$$
$$X_3 = \sqrt{6}(G_{63} + G_{27} - 2G_{41}) + \sqrt{10}(G_{25} - G_{34}).$$

Let L be the subgroup of G_2 generated by one of the above Lie subalgebras. Then L acts on $\mathbb{C}a_0$ in a natural manner. The symplectic group $SP(1)$ is the universal covering group of L and acts on $\mathbb{C}a_0$ in a natural manner. For first two cases we can express explicitly the action of $SP(1)$ on $\mathbb{C}a_0$.

Case 1. $q\cdot(x,y) = (x, qy)$, $(x,y) \in \mathbb{C}a_0$, $q \in SP(1)$.

Case 2. $q\cdot(x,y) = (qx\bar{q}, y\bar{q})$, $(x,y) \in \mathbb{C}a_0$, $q \in SP(1)$.

And for case 3 with $\theta = 0$, the action of $SP(1)$ on $\mathbb{C}a_0$ is

$$q\cdot(x,y) = (qx\bar{q}, qy\bar{q}), \quad (x,y) \in \mathbb{C}a_0, \quad q \in SP(1).$$

4. Some stable minimal cones in $\mathbb{R}^7$

In this section we determine totally real submanifolds of S^6 which are orbits of 3-dimensional closed subgroups of G_2. By Theorem 2.5 the cone over 3-dimensional totally real submanifold of S^6 is a coassociative submanifold and by Theorem 1.1 and Proposition 2.3 it is a stable minimal submanifold of $\mathbb{R}^7$. We study one by one the 4 cases of subgroups which are generated by subalgebras listed in §.3. In some cases it is convenient for us to find all orbits which are 3-dimensional minimal submanifold of S^6, since $CM-\{o\}$ is a minimal submanifold of $\mathbb{R}^7$ if and only if M is a minimal submanifold of S^6.

Case 1. In this case, $\mathbb{R}e_1$, $\mathbb{R}e_2$, $\mathbb{R}e_3$ and $\sum_{j=4}^{7} \mathbb{R}\, e_j$ are irreducible invariant subspaces. So each orbit is a small sphere or a great sphere. Only the great sphere is a minimal submanifold of S^6. So, in this case, the orbit we are looking for is a trivial one.

Case 2. This case was studied by Harvey and Lawson [4].

Theorem 4.1. *Let* L *be the subgroup of* G_2 *generated by the subalgebra spanned by* X_1, X_2 *and* X_3 *defined in* (3.2). *Then there exists exactly one orbit under* L *which is a 3-dimensional totally real submanifold of* S^6. *Namely the orbit through* $(\sqrt{5}/3)e_1 + (2/3)e_5$.

Case 3. For this case we have the following

Theorem 4.2. *Let L_θ be the subgroup of G_2 generated by the subalgebra spanned by X_1, X_2 and X_3 defined by (3.3). Then there exists exactly one orbit under L_θ which is a totally real submanifold of S^6. Namely the orbit through $(\sqrt{2}/2)(e_2 + e_5)$.*

Proof. Calculating the volume of each orbit, it is easily seen that only the orbit through $p = (\sqrt{2}/2)(e_2 + e_5)$ is a 3-dimensional minimal submanifold of S^6 under the action of L_θ on $\mathbb{C}a_0$. The tangent space of this orbit at p is spanned by

$$X_1(p) = \sqrt{2}\,(e_1 - e_6),$$

$$X_2(p) = -\sqrt{2}\cos\theta\; e_3 - \sqrt{2}\sin\theta\; e_7,$$

$$X_3(p) = \sqrt{2}\sin\theta\; e_3 - \sqrt{2}\cos\theta\; e_7.$$

Consulting the multiplication table in §.2, we get

$$J(X_1(p)) = p\cdot X_1(p) = 2\,e_4,$$

$$J(X_2(p)) = p\cdot X_2(p) = -\cos\theta\,(e_1+e_6) + \sin\theta\,(-e_2+e_5),$$

$$J(X_3(p)) = p\cdot X_3(p) = \cos\theta\,(-e_2+e_5) + \sin\theta\,(e_1+e_6).$$

So it is easily seen that the orbit is a totally real submanifold.

Q.E.D.

Case 4. For this case we have the following

Theorem 4.3. *Let L be the subgroup of G_2 generated by the Lie subalgebra spanned by X_1, X_2 and X_3 defined by (3.4). Then, under the action of L on $\mathbb{C}a_0$, there exists exactly 3-types of orbits in S^6 which are minimal submanifold in S^6 up to the action of G_2. Namely*

(i) *the orbit through* e_2, *which we denote by* M_1,
(ii) *the orbit through* e_6, *which we denote by* M_2,
(iii) *the orbit through* $(\sqrt{2}/2)(e_1 + e_3)$, *which we denote by* M_3.

The orbits M_1 and M_2 are totally real submanifolds but M_3 is not.

Proof. We omit the proof of the former part. The proof of the latter part is similar to that of Theorem 4.2.

Remark. The orbit M_1 is a space of constant curvature 1/16. This example was obtained by Ejiri [2].

REFERENCES

[1] Dynkin, E.B., Semi-simple subalgebras of semi-simple Lie algebras, Amer. Math. Soc. Transl., Ser. 2, 6(1957), 111-244.

[2] Ejiri, N., Totally real submanifold in the 6-sphere, Proc. Amer. Math. Soc., 83(1981), 759-763.

[3] Fredenthal, H., Oktaven, Ausnahmegruppen und Oktavengeometrie, Mimeographed Note, Utrecht, 1951.

[4] Harvey, R. and Lawson, H.B., A constellation of minimal varieties defined over the group G_2, Lecture Notes in Pure and Applied Math., Vol. 48, Marcel Dekker, N.Y., 1978, 1-38.

[5] Mal'cev, A.I., On semi-simple subgroups of Lie groups, Amer. Math. Soc. Transl., Ser. 1, 9(1950), 172-213.

SYMMETRIC SUBMANIFOLDS OF COMPACT SYMMETRIC SPACES

Hiro Naitoh
Department of Mathematics, Faculty of Science,
Yamaguchi University, Yamaguchi 753, Japan

In this note we study the classification of symmetric submanifolds of compact simply connected riemannian symmetric spaces. For a given symmetric submanifold S and a fixed point $p \in S$, there exists a unique totally geodesic symmetric submanifold N such that $p \in N$, $T_pN = T_pS$. To classify symmetric submanifolds S, we firstly consider the classification of totally geodesic symmetric submanifolds N (see the section 1) and next consider that of symmetric submanifolds S associated with each N. In the section 2, we reduce the classification of such S to that of certain algebraic objects (Theorem 2.3) and, in the section 3, we classify symmetric submanifolds S associated with specific totally geodesic symmetric submanifolds N by using the reduction.

0. Preliminaries

Let M be a riemannian symmetric space. A symmetric submanifold is, by definition, a connected submanifold S of M with involutive isometries t_p, $p \in S$, of M satisfying that $t_p(p) = p$, $t_p(S) = S$ and

$$(t_p)_*X = -X \quad (X \in T_pS), \quad (t_p)_*\xi = \xi \quad (\xi \in N_pS). \qquad (0.1)$$

Here t_p is called the extrinsic symmetry at p for S. A symmetric submanifold is a riemannian symmetric space with the metric induced from M. Denote by R^M the curvature tensor of M. A subspace $V \subset T_pM$ is called curvature-invariant if it satisfies that $R^M(V,V)V \subset V$. A symmetric submanifold is characterized as submanifold in the following.

Proposition 0.1. *Assume that M is simply connected. Then S is a symmetric submanifold if and only if (1) S is a complete parallel submanifold and (2) the normal spaces N_pS, $p \in S$, are curvature-invariant subspaces in T_pM.*

Proof. Denote by α the second fundamental form and by $\nabla\alpha$ its covariant derivative. Assume that S is a symmetric submanifold. Since t_p is an isometry of M, we have

$$(\nabla\alpha)((t_p)_*X,(t_p)_*Y,(t_p)_*Z) = (t_p)_*((\nabla\alpha)(X,Y,Z)) \quad (X,Y,Z \in T_pS).$$

By (0.1) it follows that $\nabla\alpha = 0$. This implies (1). Similarly, we have

$$R^M((t_p)_*\xi,(t_p)_*\zeta)(t_p)_*\eta = (t_p)_*(R^M(\xi,\zeta)\eta) \quad (\xi,\zeta,\eta \in N_pS).$$

Again by (0.1) it follows that $R^M(\xi,\zeta)\eta \in N_pS$. This implies (2).

Assume that S satisfies the conditions (1),(2). Define an isometry λ of T_pM by $\lambda(X) = -X$ ($X \in T_pS$), $\lambda(\xi) = \xi$ ($\xi \in N_pS$). By (1) and the Codazzi's equation, the subspace T_pS is curvature-invariant in T_pM. Hence, together with (2), it follows that

$$\lambda(R^M(A,B)C) = R^M(\lambda(A),\lambda(B))\lambda(C) \quad (A,B,C \in T_pM).$$

Since M is a simply connected riemannian symmetric space, there exists an isometry t_p of M such that $t_p(p) = p$, $(t_p)_* = \lambda$. By (1) geodesics $\gamma(t)$ of S are Frenet curves of M and moreover, by the uniqueness of Frenet curves for initial data, it follows that $t_p(\gamma(t)) = \gamma(-t)$ (Strübing [10]). Hence we have $t_p(S) = S$.——

Denote by $I(M)$ the Lie group of isometries of M and by $I^0(M)$ the identity component of $I(M)$. For a submanifold $L \subset M$, put $I(M,L) = \{ g \in I(M): g(L) = L \}$ and let $I^0(M,L)$ be the connected Lie subgroup of $I^0(M)$ generated by the Lie algebra of Killing vector fields of M tangent to L. If S is a symmetric submanifold, the subgroup of $I(M,S)$ generated by extrinsic symmetries t_p, $p \in S$, acts transitively on S and thus $I(M,S)$, $I^0(M,S)$ also act transitively on S.
Now two symmetric submanifolds (M,S),(M',S') are <u>equivalent</u> to each other if there exists an isometry ψ of M onto M' such that $\psi(S) = S'$.

<u>Problem.</u> Classify all the symmetric submanifolds (M,S) up to the equivalence, particularly when the ambient spaces M are simply connected.

Let (M,S) be a symmetric submanifold and L a riemannian symmetric space. Then $(M\times L,S)$ is also a symmetric submanifold. Hence we may consider the problem under the following condition (#).

(#) The submanifold S is not contained in any proper factor of riemannian product decompositions of M.

Let (M,S) be a symmetric submanifold and fix a point $p \in S$. Since T_pS is curvature-invariant, there exists a unique complete connected totally geodesic submanifold N of M such that $p \in S$, $T_pN = T_pS$. Note that $I(M,N)$ acts transitively on N and contains the extrinsic symmetry t_p. Hence (M,N) also a symmetric submanifold. We call this (M,N) the <u>totally geodesic symmetric submanifold associated with</u> (M,S). Here note that (M,N) is unique up to the equivalence without depending on the base point p.

<u>Lemma 0.2.</u> <u>Assume that</u> M <u>is simply connected.</u> <u>A symmetric submanifold</u> (M,S) <u>satisfies the condition (#) if and only if the totally geodesic symmetric submanifold</u> (M,N) <u>associated with</u> (M,S) <u>satisfies the condition (#).</u>

<u>Proof.</u> Assume that (M,N) doesn't satisfy the condition (#), i.e., that there exists a proper product decomposition $M = M_1 \times M_2$ such that $N \subset M_1$. Let $TM = TM_1 \oplus TM_2$ be the decomposition of the tangent bundle TM associated with $M = M_1 \times M_2$. Let $q \in S$. Since $I^0(M,S)$ acts transitively on S, there exists $g \in I^0(M,S) \subset I^0(M)$ such that $g(p) = q$. Note that $I^0(M) = I^0(M_1) \times I^0(M_2)$. Then it follows that $g_*(TM_1) = TM_1$. Since $T_pS = T_pN \subset T_pM_1$, it follows that $T_qS \subset T_qM_1$. Hence we have $S \subset M_1$. This implies that (M,S) doesn't satisfy the condition (#). The converse is obvious.——

Denote by $\mathcal{S}$ (resp. $\mathfrak{N}$) the set of equivalence classes of symmetric submanifolds (M,S) (resp. totally geodesic symmetric submanifolds (M,N)) satisfying the condition (#).

1. Totally geodesic symmetric submanifolds

In the following sections we always assume that M is a compact simply connected riemannian symmetric space. Let (M,N) be a totally geodesic symmetric submanifold satisfying the condition (#) and fix a base point $p \in N$. Then, (i) <u>the Lie algebra</u> $\mathfrak{g}$ <u>of</u> $I^0(M)$ <u>is compact type.</u> Denote by s_p the (intrinsic) symmetry at p of the symmetric space M. Define involutive automorphisms σ, τ of $I^0(M)$ by

$$\sigma(g) = s_p \circ g \circ s_p,\ \tau(g) = t_p \circ g \circ t_p \quad (g \in I^0(M)),$$

and also denote by the same notations their differentials. Then, (ii) σ, τ <u>are involutive automorphisms of</u> $\mathfrak{g}$ <u>commuting with each other, such that the symmetric Lie algebras</u> $(\mathfrak{g},\sigma),(\mathfrak{g},\tau)$ <u>are effective.</u> In fact, the effectivity of $(\mathfrak{g},\sigma)$ is obvious from its construction and that of $(\mathfrak{g},\tau)$ is proved by the condition (#). Let $\mathfrak{g} = \mathfrak{k} \oplus \mathfrak{p}$ be the canonical decomposition of $(\mathfrak{g},\sigma)$, i.e., $\mathfrak{k} = \{ T \in \mathfrak{g}: \sigma(T) = T \}$, $\mathfrak{p} = \{ X \in \mathfrak{g}: \sigma(X) = -X \}$. Identify T_pM with $\mathfrak{p}$. Then there exists a unique inner product $\langle,\rangle$ on $\mathfrak{g}$ whose restriction to $\mathfrak{p}$ coincides with the riemannian metric on T_pM and satisfying the following properties: (iii) σ,τ <u>leave the inner product</u> $\langle,\rangle$ <u>invariant and</u> $ad_{\mathfrak{g}}(X)$, $X \in \mathfrak{g}$, <u>are skew symmetric for</u> $\langle,\rangle$. (Note that it is proved by the commutativity of σ,τ that τ leaves $\langle,\rangle$ invariant.)

Conversely, from $(\mathfrak{g},\sigma,\tau,\langle,\rangle)$ satisfying the properties (i),(ii),(iii), we can construct a totally geodesic symmetric submanifold (M,N) satisfying the condition (#) in the following. Let G be the compact simply connected Lie group with the Lie algebra $\mathfrak{g}$ and K the connected compact Lie subgroup of G with the Lie algebra $\mathfrak{k}$. Then $M = G/K$ is a compact simply connected riemannian symmetric space with the G-invariant metric induced from $\langle,\rangle|\mathfrak{p}$. Put $o = K$ and identify T_oM with $\mathfrak{p}$. Since $\mathfrak{p}$ is τ-invariant, it is decomposed into the sum of (± 1)-eigen spaces $\mathfrak{p}_\pm$ of the restriction $\tau|\mathfrak{p}$ to $\mathfrak{p}$. Note that $\mathfrak{p}_\pm$ are curvature-invariant subspaces in $\mathfrak{p} = T_oM$. Let N be the complete connected totally geodesic submanifold of M such that $N \ni o$, $T_oN = \mathfrak{p}_-$. Since I(M,N) acts transitively on N, by the above note and proposition 0.1, N is a symmetric submanifold. The condition (#) is obvious by the effectivity of $(\mathfrak{g},\tau)$.

Now two quadruples $(\mathfrak{g},\sigma,\tau,\langle,\rangle)$, $(\mathfrak{g}',\sigma',\tau',\langle,\rangle')$ are <u>equivalent</u> to each other if there exists a Lie algebra isomorphism ϕ of $\mathfrak{g}$ onto $\mathfrak{g}'$ such that $\phi\circ\sigma = \sigma'\circ\phi$, $\phi\circ\tau = \tau'\circ\phi$, $\langle\phi(X),\phi(Y)\rangle' = \langle X,Y\rangle$ ($X,Y \in \mathfrak{g}$). Denote by $\mathcal{Y}$ the set of equivalence classes of the quadruples. The <u>direct sum</u> of some quadruples is defined naturally. For symmetric submanifolds (M_i,S_i), $i=1,\cdots,r$, put $M = M_1\times\cdots\times M_r$, $S = S_1\times\cdots\times S_r$. Then (M,S) is also a symmetric submanifold. We call this (M,S) the <u>direct product</u> of (M_i,S_i), $i=1,\cdots,r$.

<u>Theorem 1.1.</u> <u>Two correspondences between</u> $\mathfrak{N}$ <u>and</u> $\mathcal{Y}$ <u>constructed above are the inverses of each other. Moreover, under these correspondences, the direct product of totally geodesic symmetric submanifolds corresponds to the direct sum of quadruples.</u>

Let $(\mathfrak{g},\sigma,\tau)$ be a triple satisfying the properties (i),(ii) and $\mathfrak{g} = \mathfrak{k} \oplus \mathfrak{p}$ the canonical decomposition of $(\mathfrak{g},\sigma)$. Put

$$\hat{\mathfrak{g}} = \mathfrak{k} \oplus \sqrt{-1}\mathfrak{p}, \quad \hat{\tau}(X + \sqrt{-1}Y) = \tau(X) + \sqrt{-1}\tau(Y) \quad (X + \sqrt{-1}Y \in \hat{\mathfrak{g}}). \tag{1.1}$$

Then $(\hat{\mathfrak{g}},\hat{\tau})$ is an effective symmetric Lie algebra of non compact type.

Conversely, given an effective symmetric Lie algebra $(\hat{\mathfrak{g}},\hat{\tau})$ of non compact type, there exists a Cartan involution $\hat{\theta}$ of $\hat{\mathfrak{g}}$ commuting with $\hat{\tau}$ and a triple $(\mathfrak{g},\sigma,\tau)$ satisfying the properties (i),(ii) is naturally defined by the relation (1.1). We define " equivalence " and " direct sum " naturally for each object.

Corollary 1.2. These correspondences between triples $(\mathfrak{g},\sigma,\tau)$ and effective symmetric Lie algebras $(\hat{\mathfrak{g}},\hat{\tau})$ of non compact type are the inverses of each other and preserve each direct sum.

Remark. Irreducible effective symmetric Lie algebras of non compact type are classified in Berger [2] and reach to many kinds. Next note that, for a quadruple $(\mathfrak{g},\sigma,\tau,<,>)$, a decomposition of the triple $(\mathfrak{g},\sigma,\tau)$ gives rise to that of the quadruple.

Now we give an involutive correspondence of $\mathfrak{M}$ in the following.

$$\mathfrak{M} \ni (M,N) \leftrightarrow (\mathfrak{g},\sigma,\tau,<,>) \leftrightarrow (\mathfrak{g},\tau,\sigma,<,>) \leftrightarrow (M^*,N^*) \in \mathfrak{M}. \tag{1.2}$$

Denote by $\mathfrak{k}_\pm$ the (± 1)-eigen spaces of the restriction $\tau|\mathfrak{k}$ of τ to $\mathfrak{k}$. Then $\mathfrak{g} = (\mathfrak{k}_+ \oplus \mathfrak{p}_+) \oplus (\mathfrak{k}_- \oplus \mathfrak{p}_-)$ is the canonical decomposition for $(\mathfrak{g},\tau)$ and the tangent space of M^* at the base point is identified with the vector space $\mathfrak{k}_- \oplus \mathfrak{p}_-$. Under this identification, the tangent space of N^* at the base point is identified with the subspace $\mathfrak{p}_-$. Note that N and N^* are locally isometric to each other.

Remark. When M are simply connected riemannian symmetric spaces of non compact type, totally geodesic symmetric submanifolds (M,N) are got through the duality between symmetric spaces of compact type and those of non compact type.

2. Compact symmetric submanifolds

Let (M,S) be a symmetric submanifold satisfying the condition (#) and fix a base point $p \in S$. Let (M,N) be the totally geodesic symmetric submanifold associated with (M,N) and $(\mathfrak{g},\sigma,\tau,<,>)$ the quadruple corresponding to (M,S). Denote by $\mathfrak{g}_S$ the Lie algebra of $I^0(M,S)$. Since τ leaves $\mathfrak{g}_S$ invariant, $\mathfrak{g}_S$ is decomposed into the sum of the $(+1)$-eigen space $\mathfrak{k}_S$ and the (-1)-eigen space $\mathfrak{m}_S$ of the restriction $\tau|\mathfrak{g}_S$, i.e., $\mathfrak{g}_S = \mathfrak{k}_S \oplus \mathfrak{m}_S$.

Lemma 2.1. The vector space $\mathfrak{m}_s \subset \mathfrak{k}_- \oplus \mathfrak{p}_-$ is a Lie triple system (i.e., $[\mathfrak{m}_s,[\mathfrak{m}_s,\mathfrak{m}_s]] \subset \mathfrak{m}_s$) satisfying the following. (a) $[\mathfrak{m}_s,\mathfrak{m}_s] \subset \mathfrak{k}_+$. (b) the orthogonal projection: $\mathfrak{m}_s \to \mathfrak{p}_-$ is a linear isomorphism.

Proof. It is obvious by the definition of $\mathfrak{m}_s$ that $\mathfrak{m}_s$ is a Lie triple system in $\mathfrak{k}_- \oplus \mathfrak{p}_-$. Firstly we show that $\mathfrak{k}_s = \mathfrak{g}_s \cap \mathfrak{k}$. Let $T \in \mathfrak{k}_s$. Since $\tau(T) = T$, it follows that $t_p \circ e^{tT} \circ t_p = e^{tT}$ for $t \in \mathbb{R}$, where $e^{(*)}$ denotes the exponential map. Hence $e^{tT}(p)$, $t \in \mathbb{R}$, are fixed points of t_p. Since $e^{tT}(p) \in S$, it follows that $e^{tT}(p) = p$ and thus $T \in \mathfrak{k}$. Conversely let $T \in \mathfrak{g}_s \cap \mathfrak{k}$. Since $e^{tT}(p) = p$ ($t \in \mathbb{R}$), it follows that $e^{tT}(p) = (t_p \circ e^{tT} \circ t_p)(p) = p$. Note that $(e^{tT})_* T_pS = T_pS$, $(e^{tT})_* N_pS = N_pS$. Then we have $(e^{tT})_{*p} = (t_p \circ e^{tT} \circ t_p)_{*p}$. Hence it follows that $e^{tT} = t_p \circ e^{tT} \circ t_p$. This implies that $\tau(T) = T$ and thus $T \in \mathfrak{k}_s$.

The claim (a) is obvious since $\mathfrak{k}_s \subset \mathfrak{k} \cap (\mathfrak{k}_+ \oplus \mathfrak{p}_+) = \mathfrak{k}_+$. We show the claim (b). Let $T \in \mathfrak{m}_s \cap \mathfrak{k}_-$. Then it follows that $T \in \mathfrak{g}_s \cap \mathfrak{k} = \mathfrak{k}_s \subset \mathfrak{k}_+$ and thus $T = 0$. This implies that the projection is injective. Since $\dim \mathfrak{m}_s = \dim \mathfrak{p}_-$, the projection is bijective.——

Conversely, from $(\mathfrak{g},\sigma,\tau,<,>:\mathfrak{m})$ with a vector space $\mathfrak{m}$ satisfying the properties in Lemma 2.1, a symmetric submanifold (M,S) satisfying the condition (#) is constructed in the following. Let (M,N) be the totally geodesic symmetric submanifold corresponding to the quadruple $(\mathfrak{g},\sigma,\tau,<,>)$ and let G,K,o be the same notations as given in the section 1. Put $\mathfrak{g}_{\mathfrak{m}} = [\mathfrak{m},\mathfrak{m}] \oplus \mathfrak{m}$. Since $\mathfrak{m}$ is a Lie triple system, $\mathfrak{g}_{\mathfrak{m}}$ is a Lie subalgebra of $\mathfrak{g}$. Let $G_{\mathfrak{m}}$ be the connected Lie subgroup of G with the Lie algebra $\mathfrak{g}_{\mathfrak{m}}$ and S the orbit of o by $G_{\mathfrak{m}}$.

Lemma 2.2. The submanifold (M,S) is a symmetric submanifold satisfying the condition (#) and associated with (M,N).

Proof. To prove that (M,S) is a symmetric submanifold, we prove Proposition 0.1,(1),(2). Since S is $G_{\mathfrak{m}}$-equivalent, we may show them at the base point o. Note that $T_oS = T_oN \subset T_oM$ by the property (b). This implies (2). Since T_oN is a curvature-invariant subspace in T_oM, by the Codazzi's equation, the trilinear form $(\nabla\alpha)_o$ is symmetric. To prove (1), we may show that $(\nabla\alpha)_o(X,X,X) = 0$ for $X \in T_oS$. Identify T_oS, N_oS with $\mathfrak{m}$, $\mathfrak{p}_+$ respectively. Then, by using the property (a), we can calculate $(\nabla\alpha)_o$ as follows: for $X = X_{\mathfrak{k}} + X_{\mathfrak{p}} \in \mathfrak{m}$ ($X_{\mathfrak{k}} \in \mathfrak{k}_-$, $X_{\mathfrak{p}} \in \mathfrak{p}_-$),

$$(\nabla\alpha)_o(X,X,X) = \text{the } \mathfrak{p}_+\text{-component of } [X_{\mathfrak{k}},[X_{\mathfrak{k}},X_{\mathfrak{p}}]] = 0.$$

Hence it follows that $(\nabla\alpha)_o = 0$. The other claims are obvious.——

Now two $(\mathfrak{g},\sigma,\tau,<,>:\mathfrak{m}),(\mathfrak{g}',\sigma',\tau',<,>':\mathfrak{m}')$ are <u>equivalent</u> to each other if there exists an isomorphism ϕ of $(\mathfrak{g},\sigma,\tau,<,>)$ onto $(\mathfrak{g}',\sigma',\tau',<,>')$ such that $\phi(\mathfrak{m}) = \mathfrak{m}'$. Denote by $\mathfrak{M}$ the set of equivalence classes and moreover define the " direct sum " of objects in $\mathfrak{M}$ naturally.

<u>Theorem 2.3. Two correspondences between $\mathcal{S}$ and $\mathfrak{M}$ constructed above are the inverses of each other. Moreover, under these correspondences, the direct product of symmetric submanifolds corresponds to the direct sum of objects in $\mathfrak{M}$.</u>

Let $(\mathfrak{g},\sigma,\tau,<,>:\mathfrak{m})\in\mathfrak{M}$. By the property (b), there exists a unique linear map λ of $\mathfrak{p}_-$ to $\mathfrak{k}_-$ such that

$$\mathfrak{m} = \{ X+\lambda(X)\in \mathfrak{k}_-\oplus \mathfrak{p}_-: X\in \mathfrak{p}_- \}. \tag{2.1}$$

This λ satisfies the following conditions (A),(L). For $X,Y,Z\in\mathfrak{p}_-$,

(A) $[X,\lambda(Y)] = [Y,\lambda(X)]$,

(L) $\lambda([[X,Y],Z]) + \lambda([[\lambda(X),\lambda(Y)],Z] =$
$[[X,Y],\lambda(Z)] + [[\lambda(X),\lambda(Y)],\lambda(Z)]$.

In fact, the condition (A) follows by the condition (a) for $\mathfrak{m}$ and the condition (L) follows by the fact that $\mathfrak{m}$ is a Lie triple system.

Conversely, given a linear map λ of $\mathfrak{p}_-$ to $\mathfrak{k}_-$ satisfying the conditions (A),(L), we can construct a vector space $\mathfrak{m}$ satisfying the properties in Lemma 2.2 by the relation (2.1).

Two $(\mathfrak{g},\sigma,\tau,<,>:\lambda),(\mathfrak{g}',\sigma',\tau',<,>':\lambda')$ are <u>equivalent</u> to each other if there exists an isomorphism ϕ of $(\mathfrak{g},\sigma,\tau,<,>)$ onto $(\mathfrak{g}',\sigma',\tau',<,>')$ such that $\phi\circ\lambda = \lambda'\circ\phi$. Denote by $\mathfrak{A}$ the set of equivalence classes and moreover define the " direct sum " of objects in $\mathfrak{A}$ naturally.

<u>Corollary 2.4. The set $\mathfrak{M}$ corresponds to the set $\mathfrak{A}$ under the above correspondences. Moreover these correspondences preserve each direct sum.</u>

<u>Remark.</u> Let (M,S) be a symmetric submanifold satisfying the condition (#) and $(\mathfrak{g},\sigma,\tau,<,>:\mathfrak{m})$ the object corresponding to (M,S). Let

(M^*,N^*) be the totally geodesic symmetric submanifold constructed by (1.2) from the associated totally geodesic symmetric submanifold (M,N). Since $\mathfrak{m}$ is a Lie triple system in $\mathfrak{k}_- \oplus \mathfrak{p}_-$, it is curvature-invariant in the tangent space $T_{o^*}M^*$ at the base point o^*. Hence there exists a unique complete connected totally geodesic submanifold S^* such that $S^* \ni o^*$, $T_{o^*}S^* = \mathfrak{m}$. To seek (M,S) we may find out S^* satisfying the properties (a),(b) at o^*. Generally, many examples of totally geodesic submanifolds are known in Nagano-Chen [3]. Note that S,S^* are locally isomorphic as symmetric space, but that they are not always isometric even locally.

3. Specific compact symmetric submanifolds

When M are compact simply connected riemannian symmetric spaces of rank one, totally geodesic symmetric submanifolds (M,N) are the following seven kinds: (S^n,S^r), $(\mathbb{C}P^n,\mathbb{C}P^r)$, $(\mathbb{C}P^n,\mathbb{R}P^n)$, $(\mathbb{H}P^n,\mathbb{H}P^r)$, $(\mathbb{H}P^n,\mathbb{C}P^n)$, $(CayP^2,S^8)$, $(CayP^2,\mathbb{H}^2P)$. Here S^ℓ denotes an ℓ-dimensional sphere and $\mathbb{R}P^\ell$, $\mathbb{C}P^\ell$, $\mathbb{H}P^\ell$ ℓ-dimensional projective spaces over the real, complex, quaternion number fields respectively and $CayP^2$ the Cayley projective plane. The symmetric submanifolds (M,S) associated with each (M,N) are completely classified in [1],[4],[11],[8],[12],[9],[5],[13],[14], etc. Now we describe two results got by applying Theorem 2.3. Let (M,N) be a totally geodesic symmetric submanifold satisfying the condition (#) and $(\mathfrak{g},\sigma,\tau,\langle,\rangle)$ the quadruple corresponding to (M,N). They are called <u>irreducible</u> if they are not properly decomposed into the direct product or direct sum of objects.

<u>Case I.</u> <u>Assume that</u> $(\mathfrak{g},\sigma,\tau,\langle,\rangle)$ <u>is an irreducible quadruple such that</u>

$$C(\mathfrak{k}_+ \oplus \mathfrak{p}_+) \cap \mathfrak{p}_+ \neq \{0\}, \tag{3.1}$$

<u>where</u> $C(*)$ <u>denotes the center of</u> $*$. In this case the dimension of $C(\mathfrak{k}_+ \oplus \mathfrak{p}_+) \cap \mathfrak{p}_+$ equals one and, for some H_0 in it, $J = ad(H_0)$ gives a complex structure on $\mathfrak{k}_- \oplus \mathfrak{p}_-$. Hence M^* is a hermitian symmetric space and N^* is a totally real submanifold such that $\dim_{\mathbb{C}} M^* = \dim_{\mathbb{R}} N^*$. In fact, (M^*,N^*) exhausts all the pairs of irreducible symmetric R-spaces and the complexificated hermitian symmetric spaces. We list up $(\hat{\mathfrak{g}},\hat{\tau})$, $(\mathfrak{g},\sigma)$, $(\mathfrak{g},\tau)$, $\mathfrak{k}_+$, $\mathfrak{k}_-$, $\mathfrak{p}_+$, $\mathfrak{p}_-$ for this case in the following table.

TABLE

No.	$(\hat{\mathfrak{g}},\hat{\tau})$	$(\mathfrak{g},\sigma)$	$(\mathfrak{g},\tau)$	$\mathfrak{k}_+$
1	$sl(n;\mathbb{C})/sl(i;\mathbb{C})\oplus sl(n-i;\mathbb{C})\oplus\mathbb{C}^*$	$su(n)$	$su(n)/s(u(i)\oplus u(n-i)) \oplus su(n)/s(u(i)\oplus u(n-i))$	$s(u(i)\oplus u(n-i))$
2	$so(2n;\mathbb{C})/sl(n;\mathbb{C})\oplus\mathbb{C}^*$	$so(2n)$	$so(2n)/u(n) \oplus so(2n)/u(n)$	$u(n)$
3	$so(n;\mathbb{C})/so(n-2;\mathbb{C})\oplus\mathbb{C}^*$	$so(n)$	$so(n)/so(n-2)\oplus T \oplus so(n)/so(n-2)\oplus T$	$so(n-2)\oplus T$
4	$sp(n;\mathbb{C})/sl(n;\mathbb{C})\oplus\mathbb{C}^*$	$sp(n)$	$sp(n)/u(n) \oplus sp(n)/u(n)$	$u(n)$
5	$E_6^{\mathbb{C}}/so(10;\mathbb{C})\oplus\mathbb{C}^*$	E_6	$E_6/so(10)\oplus T \oplus E_6/so(10)\oplus T$	$so(10)\oplus T$
6	$E_7^{\mathbb{C}}/E_6^{\mathbb{C}}\oplus\mathbb{C}^*$	E_7	$E_7/E_6\oplus T \oplus E_7/E_6\oplus T$	$E_6\oplus T$
7	$su(n,n)/sl(n;\mathbb{C})\oplus\mathbb{R}$	$su(2n)/s(u(n)\oplus u(n))$	$su(2n)/s(u(n)\oplus u(n))$	$su(n)$
8	$so^*(4n)/su^*(2n)\oplus\mathbb{R}$	$so(4n)/u(2n)$	$so(4n)/u(2n)$	$sp(n)$
9	$sp(n;\mathbb{R})/sl(n;\mathbb{R})\oplus\mathbb{R}$	$sp(n)/u(n)$	$sp(n)/u(n)$	$so(n)$
10	$E_7^3/E_6^4\oplus\mathbb{R}$	$E_7/E_6\oplus T$	$E_7/E_6\oplus T$	F_4
11	$sl(n;\mathbb{R})/sl(i;\mathbb{R})\oplus sl(n-i;\mathbb{R})\oplus\mathbb{R}$	$su(n)/so(n)$	$su(n)/s(u(i)\oplus u(n-i))$	$so(i)\oplus so(n-i)$
12	$su^*(2n)/su^*(2i)\oplus su^*(2n-2i)\oplus\mathbb{R}$	$su(2n)/sp(n)$	$su(2n)/s(u(2i)\oplus u(2n-2i))$	$sp(i)\oplus sp(n-i)$
13	$so(i,n-i)/so(i-1,n-i-1)\oplus\mathbb{R}$	$so(n)/so(i)\oplus so(n-i)$	$so(n)/so(n-2)\oplus T$	$so(i-1)\oplus so(n-i-1)$
14	$so(n,n)/sl(n;\mathbb{R})\oplus\mathbb{R}$	$so(2n)/so(n)\oplus so(n)$	$so(2n)/u(n)$	$so(n)$
15	$sp(n,n)/su^*(2n)\oplus\mathbb{R}$	$sp(2n)/sp(n)\oplus sp(n)$	$sp(2n)/u(2n)$	$sp(n)$
16	$E_6^1/so(5,5)\oplus\mathbb{R}$	$E_6/sp(4)$	$E_6/so(10)\oplus T$	$so(5)\oplus so(5)$
17	$E_6^4/so(1,9)\oplus\mathbb{R}$	E_6/F_4	$E_6/so(10)\oplus T$	$so(9)$
18	$E_7^1/E_6^1\oplus\mathbb{R}$	$E_7/su(8)$	$E_7/E_6\oplus T$	$sp(4)$

No.	$\mathfrak{k}_-$, $\mathfrak{p}_-$	$\mathfrak{p}_+$	Remark
1	$su(n)/s(u(i)\oplus u(n-i))$	$s(u(i)\oplus u(n-i))$	$1\le i\le n-i$
2	$so(2n)/u(n)$	$u(n)$	$n\ge 5$
3	$so(n)/so(n-2)\oplus T$	$so(n-2)\oplus T$	$n\ge 3,\ \ne 4$
4	$sp(n)/u(n)$	$u(n)$	$n\ge 3$
5	$E_6/so(10)\oplus T$	$so(10)\oplus T$	
6	$E_7/E_6\oplus T$	$E_6\oplus T$	
7	$u(n)$	$u(n)$	$n\ge 2$
8	$T\oplus su(2n)/sp(n)$	$T\oplus su(2n)/sp(n)$	$n\ge 3$
9	$T\oplus su(n)/so(n)$	$T\oplus su(n)/so(n)$	$n\ge 3$

10	$T \oplus E_6/F_4$	$T \oplus E_6/F_4$	
11	$so(n)/so(i) \oplus so(n-i)$	$T \oplus su(i)/so(i) \oplus su(n-i)/so(n-i)$	$1 \leq i \leq n-i$
12	$sp(n)/sp(i) \oplus sp(n-i)$	$T \oplus su(2i)/sp(i) \oplus su(2n-2i)/sp(n-i)$	$1 \leq i \leq n-i$
13	$so(i)/so(i-1) \oplus so(n-i)/so(n-i-1)$	$T \oplus so(n-2)/so(i-1) \oplus so(n-i-1)$	$i=1$: $n-i \geq 4$, $\neq 5$. $i=2$: $n-i \geq 3$, $\neq 4$. $n-i \geq i \geq 3$ (except $i=n-i=3$).
14	$so(n)$	$T \oplus su(n)/so(n)$	$n \geq 5$
15	$sp(n)$	$T \oplus su(2n)/sp(n)$	$n \geq 2$
16	$sp(4)/sp(2) \oplus sp(2)$	$T \oplus so(10)/so(5) \oplus so(5)$	
17	$F_4/so(9)$	$T \oplus so(10)/so(9)$	
18	$su(8)/sp(4)$	$T \oplus E_6/sp(4)$	

Now define a linear map λ_c of $\mathfrak{p}_-$ to $\mathfrak{k}_-$ for $c \in \mathbb{R}$ by $\lambda_c = c \cdot J|\mathfrak{p}_-$ and put $\mathfrak{m}_c = \{ X + cJ(X) \colon X \in \mathfrak{p}_- \}$. Then λ_c satisfies the conditions (A),(L). Let (M,S_c) be the symmetric submanifold corresponding to the object $(\mathfrak{g},\sigma,\tau,\langle,\rangle:\lambda_c)$.

<u>Lemma 3.1</u>. (1) S_c <u>is a pseudo-umbilical submanifold</u>. (2) <u>Two symmetric submanifolds</u> $(M,S_c),(M,S_{-c})$ <u>are equivalent to each other</u>. <u>And symmetric submanifolds</u> (M,S_c), $c \geq 0$, <u>are not equivalent to each other</u>.

<u>Proof.</u> (1) Identify T_oS_c, T_oN with $\mathfrak{m}_c$, $\mathfrak{p}_-$ respectively and moreover identify $\mathfrak{m}_c$ with $\mathfrak{p}_-$ by the correspondence: $\mathfrak{m}_c \ni X + cJ(X) \leftrightarrow X \in \mathfrak{p}_-$. And also identify $N_oS_c = N_oN$ with $\mathfrak{p}_+$. Denote by R^c, R^M the curvature tensors of S_c, M respectively and by α^c the second fundamental form of S_c. Then we can show that

$$R^c(X,Y)Z = (1+c^2)R^M(X,Y)Z = -(1+c^2)[[X,Y],Z] \tag{3.2}$$

and

$$\alpha^c(X,Y) = [cJ(X),Y] \quad (X,Y,Z \in \mathfrak{p}_-). \tag{3.3}$$

Denote by H^c the mean curvature vector field of S_c and by R^*,Ric^* the curvature tensor, the Ricci tensor of M^* respectively. Let $e_1, \cdots, e_\ell$ be an orthonormal basis of $\mathfrak{p}_-$. Then, by the Gauss' equation for S_c and (3.2),(3.3), it follows that

$$\begin{aligned} \ell \langle \alpha^c(X,Y),H^c \rangle &= \Sigma c^2 \langle R^M(X,e_i)e_i,Y \rangle + \Sigma \langle [cJ(X),e_i],[cJ(Y),e_i] \rangle \\ &= c^2\{\Sigma \langle R^*(X,e_i)e_i,Y \rangle + \Sigma \langle R^*(X,Je_i)Je_i,Y \rangle\} \\ &= c^2 Ric^*(X,Y) \end{aligned}$$

for $X,Y \in \mathfrak{p}_-$. Note that M^* is Einstein. Hence S_c is pseudo-umbilical. (2) The first claim follows by the fact that the isomorphism σ of $(\mathfrak{g},\sigma,\tau,<,>)$ satisfies that $\sigma(\mathfrak{m}_c) = \mathfrak{m}_{-c}$. The second claim is obvious by (3.2).——

Conversely, let λ be a linear map of $\mathfrak{p}_-$ to $\mathfrak{k}_-$ satisfying the conditions (A),(L) and define a linear subspace $\mathfrak{m}$ in $\mathfrak{k}_- \oplus \mathfrak{p}_-$ by (2.1). Put $\tilde{\lambda} = J\circ\lambda \in \mathrm{End}(\mathfrak{p}_-)$.

<u>Lemma 3.2.</u> $\mathrm{ad}_{\mathfrak{p}_-}([S,T])\cdot\tilde{\lambda} = 0$ <u>for</u> $S,T \in \mathfrak{p}_+$.

<u>Proof.</u> By the condition (A), it follows that $<[T,X+\lambda(X)],Y+\lambda(Y)> = <T,[X,\lambda(Y)]+[\lambda(X),Y]> = 0$ for $T \in \mathfrak{p}_+$, $X,Y \in \mathfrak{p}_-$ and thus $[\mathfrak{p}_+,\mathfrak{m}] \subset \mathfrak{m}^\perp$, where $\mathfrak{m}^\perp$ denotes the orthogonal complement of $\mathfrak{m}$ in $\mathfrak{k}_- \oplus \mathfrak{p}_-$. Particularly we have $J\mathfrak{m} = \mathfrak{m}^\perp$ and thus $[\mathfrak{p}_+,\mathfrak{m}] = J\mathfrak{m}$. Hence, for $X \in \mathfrak{p}_-$, there exists $Y \in \mathfrak{p}_-$ such that $[T,X+\lambda(X)] = JY + J\lambda(Y)$, i.e., $[T,X] = JY$ and $[T,\lambda(X)] = J\lambda(Y)$. Canceling Y, we have $J[T,\lambda(X)] = \lambda(J[T,X])$. This implies that $J\circ\mathrm{ad}(T)\circ\lambda = \lambda\circ J\circ\mathrm{ad}(T)$, i.e., $\mathrm{ad}(T)\circ\tilde{\lambda} = -J\circ\tilde{\lambda}\circ J\circ\mathrm{ad}(T)$ on $\mathfrak{p}_-$. Hence it follows that $\mathrm{ad}(S)\circ\mathrm{ad}(T)\circ\tilde{\lambda} = \tilde{\lambda}\circ\mathrm{ad}(S)\circ\mathrm{ad}(T)$ and thus $\mathrm{ad}([S,T])\circ\tilde{\lambda} = \tilde{\lambda}\circ\mathrm{ad}([S,T])$ for $S,T \in \mathfrak{p}_+$.——

Let $(\mathbb{C}Q^n,S^n)$ ($n \geq 1$) be the totally real totally geodesic symmetric submanifold S^n of the n-dimensional complex quadric $\mathbb{C}Q^n$ which appears as (M^*,N^*) in the case when No.13, $i = 1$ in the table.

<u>Theorem 3.3.</u> <u>Assume that</u> (M^*,N^*) <u>is not equivalent to any</u> $(\mathbb{C}Q^n,S^n)$, $n \geq 1$. <u>Then the symmetric submanifold</u> (M,S) <u>corresponding to</u> $(\mathfrak{g},\sigma,\tau,<,>:\lambda)$ <u>is equivalent to some</u> (M,S_c).

<u>Outline of the proof.</u> Decide $\tilde{\lambda}$ satisfying Lemma 3.2 and moreover, in them, find out $\tilde{\lambda}$ such that λ satisfies the conditions(A),(L).——

<u>Remark.</u> When (M^*,N^*) is equivalent to $(\mathbb{C}Q^n,S^n)$, (M,N) is equivalent to (S^{n+1},S^n) and thus (M,S) is a <u>generalized Clifford torus</u> of S^{n+1} from the classification of symmetric submanifolds of S^{n+1}. In this case it follows that $\mathfrak{p}_+ = \mathbf{T}$ and thus $[\mathfrak{p}_+,\mathfrak{p}_+] = \{0\}$. Hence Lemma 3.2 restricts nothing for $\tilde{\lambda}$.

<u>Remark.</u> The symmetric submanifolds (M,S_c) associated with (M,N) in No.1$\sim$ No.6 were constructed in Tsukada [15].

Case II. Let (M,N) be a totally geodesic symmetric submanifold satisfying the condition (#) and $(\mathfrak{g},\sigma,\tau,<,>)$ the quadruple corresponding to (M,N). Let λ be a linear map of $\mathfrak{p}_-$ to $\mathfrak{k}_-$ satisfying the conditions (A),(L).

Lemma 3.4. The kernel of λ is an $\mathrm{ad}_{\mathfrak{p}_-}([\mathfrak{p}_-,\mathfrak{p}_-])$-submodule in $\mathfrak{p}_-$.

Proof. Let $X,Y,Z \in \mathfrak{p}_-$. We may show that $\lambda([[X,Y],Z]) = 0$ if $\lambda(Z) = 0$. Assume that $\lambda(Z) = 0$. By the condition (L), it follows that

$$\lambda([[X,Y],Z]) + \lambda([[\lambda(X),\lambda(Y)],Z]) = 0$$

and moreover, by the condition (A), it follows that

$$\begin{aligned}[[\lambda(X),\lambda(Y)],Z] &= -[[\lambda(Y),Z],\lambda(X)] - [[Z,\lambda(X)],\lambda(Y)] \\ &= -[[\lambda(Z),Y],\lambda(X)] - [[X,\lambda(Z)],\lambda(Y)] = 0.\end{aligned}$$

Hence we have $\lambda([[X,Y],Z])= 0$.——

Theorem 3.5. Let (M,N) be a totally geodesic symmetric submanifold satisfying the condition (#) such that (1) $\mathfrak{p}_-$ is irreducible as $\mathrm{ad}_{\mathfrak{p}_-}([\mathfrak{p}_-,\mathfrak{p}_-])$-module and (2) $\dim \mathfrak{p}_- > \dim \mathfrak{k}_-$. Then a symmetric submanifold (M,S) associated with (M,N) is equivalent to (M,N).

Proof. By Lemma 3.4 and (1), the kernel of λ is $\{0\}$ or $\mathfrak{p}_-$. Since λ is not injective by (2), it follows that $\lambda = 0$. This implies that $(M,S) = (M,N)$.——

Among the irreducible totally geodesic symmetric submanifolds, there exist many examples satisfying the conditions in Theorem 3.5. We introduce one of them.

Example: $(M,N) = (\mathbb{C}Q^n,S^n)$ ($n \geq 2$). In this case, since (M^*,N^*) is equivalent to (S^{n+1},S^n), it is obvious that the conditions in Theorem 3.5 are satisfied.

REFERENCES

[1] E.Backes-H.Reckziegel: On symmetric submanifolds of spaces of constant curvature, Math. Ann. 263(1983), 419-433.

[2] M.Berger: Les espaces symetriques non compacts, Ann. Sci. Ecole Norm. Sup. (4), 74(1957), 85-177.

[3] B.Y.Chen-T.Nagano: Totally geodesic submanifolds of symmetric spaces II, Duke Math. J., 45(1978), 405-425.

[4] D.Ferus: Symmetric submanifolds of euclidean space, Math. Ann., 247(1980), 81-93.

[5] S.Funabashi: Totally complex submanifolds of a quaternionic Kaehlerian manifold, Kodai Math. J., 2(1979), 314-336.

[6] S.Helgason: Differential Geometry, Lie groups and Symmetric spaces, Academic Press, New York, 1978.

[7] S.Kobayashi-K.Nomizu: Foundations of Differential Geometry I,II, Interscience, New York, 1963, 1969.

[8] H.Nakagawa-R.Takagi: On locally symmetric Kaehler submanifolds in a complex projective space, J. Math. Soc. Japan, 28(1976),638-667.

[9] H.Naitoh-M.Takeuchi: Totally real submanifolds and symmetric bounded domains, Osaka J. Math., 19(1982), 717-731.

[10] W.Strübing: Symmetric submanifolds of riemannian manifolds, Math. Ann., 245(1979),37-44.

[11] M.Takeuchi: Parallel submanifolds of space forms, Manifolds and Lie groups, in honor of Y.Matsushima, ed. by J.Hano et al., Birkhäuser, Boston, 1981, 429-447.

[12] ————: Parallel projective manifolds and symmetric bounded domains, Preprint.

[13] K.Tsukada: Parallel submanifolds in a quaternion projective space, Preprint.

[14] ————: Parallel submanifolds of Cayley plane, Preprint.

[15] ————: Parallel Kaehler submanifolds of hermitian symmetric spaces, Preprint.

APPENDIX

GAUSS MAPS OF SURFACES WITH CONSTANT MEAN CURVATURE

Katsuei Kenmotsu*)
Department of Mathematics, College of
General Education, Tôhoku University
Kawauchi, Sendai 980/ Japan

We hope to survey a recent progress about the theory of images of Gauss maps of surfaces with constant mean curvature.

1. Gauss maps of minimal surfaces. Let M be a complete, oriented minimal surface in the 3-dimensional Euclidean space R^3 and g the Gauss map of M which can be considered as a holomorphic mapping from M to the Riemann sphere S^2.

The famous theorem due to Xavier [18] proves that the number of points in $S^2-g(M)$ is smaller than or equal to six if M is not a plane. (cf. For the proof, see also [12]).

As the historical background of this result, we are going to mention two theorems by S.Bernstein and R.Osserman [14]:

A) Any non-parametric minimal surface in R^3 defined on R^2 must be a plane ; and

B) The closure of g(M) in S^2 is equal to S^2 if M is not a plane.

It is known by K.Voss [17] that for each d, $0 \leq d \leq 4$, there exists a complete minimal surface M in R^3 such that the number of points in $S^2-g(M)$ is equal to d. We shall remark that there is no result for d = 5 or 6.

2. Non-parametric surface with constant mean curvature. It is desirable to construct a rich theory for Gauss maps of surfaces with non-zero constant mean curvature. At first we shall study complete non-parametric surfaces with constant mean curvature.

S.Bernstein proved in 1912 that there is no non-parametric surface in R^3 which is defined on R^2 and has non-zero constant mean curvature.

---

*) The following is the text of my talking which I delivered in a conference held in the University of Tsukuba, Japan February, 1984.

In 1955, Heinz [7] proved the following generalization by giving very simple proof: Let $z = f(x,y)$, $x^2+y^2 \leq r^2$, be a non-parametric surface in R^3 and H its mean curvature function. If $|H| \geq c > 0$ for some constant c, then $r \leq 1/c$.

Later, S.S.Chern [3] has obtained an n-dimensional generalization of the theorem cited above and, in that paper, he proposed a problem:

Study complete, oriented, surfaces with constant mean curvature in R^3 such that the Gauss map, say g, omits an open set of the unit sphere S^2.

3. Examples. For the problem of S.S.Chern we wish to know as many as examples as possible.

1) $S^2(r)$ denotes the 2-sphere with the radius r centered 0 of R^3, which is a closed umbilical surface with $H = 1/r$ and it is clear that $g(S^2(r)) = S^2$.
2) R^2 denotes the 2-plane in R^3 which is a complete totally geodesic surface with $H \equiv 0$ and $g(R^2) = \{1 \text{ point}\}$.
3) $R^1 x S^1(r)$ denotes the right circular cylinder which has constant principal curvatures 0 and $1/r$, and $g(R^1 x S^1(r))$ = a great circle in S^2.
4) Unduloid (cf. Delaunay[4] and Kenmotsu[11]). It is an imbedded surface of revolution with constant mean curvature($\neq 0$), but not constant principal curvatures and whose image of the Gauss map lies in an arbitrary thin symmetric band about an equator.
5) Nodoid (cf. [4], [11]). It is an immersed surface of revolution with constant mean curvature, but not constant principal curvatures and the image of the Gauss map covers S^2 infinitely many times.
6) Helicoidal surface with H=constant. In [5], do Carmo and Dajczer studied helicoidal surfaces with non-zero constant mean curvature. Recently by making use of this result, W.Seaman [16] has proved that there exists a complete helicoidal surface of constant mean curvature in R^3 such that the image of the Gauss map is a symmetric band about an equator.
7) Lawson's surface. In [13] Lawson constructed a complete imbedded surface with constant mean curvature($\neq 0$) which lie between two parallel planes.

These constructions of examples are very concrete. Now we are going to talk about general methods of existence for such surfaces.

A) The Plateau problem for H-surfaces has been treated by many papers.

As a reference, we cite a paper by S.Hildebrandt [8].

B) Lawson [13] has proved the following: (M,ds^2) denotes a complete simply connected 2-dimensional Riemannian manifold and we assume that $M \longrightarrow S^3(1)$ is an isometric minimal immersion of M into the 3-dimensional unit sphere. Then the Gaussian curvature K of M is always smaller than or equal to one. If $K < 1$, then we have a one parameter family of isometric immersions with constant mean curvature($\neq 0$) of M into R^3.

Since we know many minimal surfaces in S^3, there are also a lot of surfaces with constant mean curvature ($\neq 0$) in R^3 by this correspondence.

C) It is known by Ruh-Vilms [15] that the Gauss map of a surface with constant mean curvature is harmonic.

Conversely, in 1979, we have shown [10] that given any harmonic, nowhere $\pm$ holomorphic map g of a Riemann surface M into S^2, there is a surface of constant mean curvature $H \neq 0$ in R^3, unique up to translation, which has the Gauss map, in stereographic coordinates, g.

4. Known results and problems. In 1982, Hoffman, Osserman and Schoen [9] proved that the image of the Gauss map of a complete surface of constant mean curvature in R^3 cannot lie in a closed hemisphere unless it is a plane or a right circular cylinder.

Their proof uses essentially a lemma by Fischer-Colbrie and Schoen [6] and the uniformization theorem of Riemann surfaces.

By an observation of known examples described in the section 3, do Carmo conjectured that the image of the Gauss map of a complete surface with constant mean curvature($\neq 0$) in R^3 contained an equator of S^2 if it is not a plane.

Some generalization to the higher dimensional case of the theorem by Hoffman, Osserman and Schoen cited above may contain difficult points, because there are non-trivial complete minimal graphs in R^{n+1}, $n \geq 8$, by Bombieri, de Giorgi and Giusti [1].

But, S.Y.Cheng and S.T.Yau [2] have shown that if M^n is a complete convex hypersurface with constant mean curvature in R^{n+1}, then M^n must be a generalized cylinder. In their proof, the convexity of M^n is used to show that the image of the Gauss map is contained in a closed hemisphere of the unit n-sphere.

Therefore we conjecture that if the image of the Gauss map of a complete oriented hypersurface with constant mean curvature in R^{n+1} is contained in a closed hemisphere, then M^n is a minimal hypersurface or a generalized cylinder.

References

[1] E.Bombieri, E.DeGiorgi and E.Giusti, Minimal cones and the Bernstein Problem, Inv.math., 7(1969), 243-268.

[2] S.Y.Cheng and S.T.Yau, Differential equations on Riemannian manifolds and their geometric applications, Comm.Pure and App.Math., 28(1975), 333-354.

[3] S.S.Chern, On the curvature of a piece of hypersurface in Euclidean space, Abh.Math.sem.univ.Hamburg, 29(1965), 77-91.

[4] C.Delaunay, Sur la surface de révolution dont la courbure moyenne est constante, J.Math.Pures.Appl., Ser.1. 6(1841), 309-320.

[5] M.doCarmo and M.Dajczer, Helicoidal surfaces with constant mean curvature, Tôhoku Math.J., 34(1982), 425-436.

[6] D.Fischer-Colbrie and R.Schoen, The structure of complete stable minimal surfaces in 3-manifolds of nonnegative scalar curvature, Comm.Pure and App.Math., 33(1980), 199-211.

[7] E.Heinz, Über Flachen mit eineindeutiger Projektion auf eine Ebene, deren Krümmungen durch Ungleichungen eingeschränkt sind, Math.Ann., 129(1955), 451-454.

[8] S.Hildebrandt, On the Plateau problem for surfaces of constant mean curvature, Comm.Pure and App.Math., 23(1970), 97-114.

[9] D.A.Hoffman, R.Osserman and R.Schoen, On the Gauss map of complete surfaces of constant mean curvature in R^3 and R^4, Comm.Math.Helv., 57(1982), 519-531.

[10] K.Kenmotsu, Weierstrass formula for surfaces of prescribed mean curvature, Math.Ann., 245(1979), 89-99.

[11] ________, Surfaces of revolution with prescribed mean curvature, Tôhoku Math.J., 32(1980), 147-153.

[12] ________, Introduction to the theory of classical minimal surfaces, Surveys in Geometry 1981/1982, Part 1 (in Japanese).

[13] B.Lawson, Complete minimal surfaces in S^3, Ann. Math., 92(1970), 335-374.

[14] R.Osserman, Proof of a conjecture of Nirenberg, Comm.Pure and App.Math., 12(1959), 229-232.

[15] E.A.Ruh and J.Vilms, The tension field of the Gauss map, Trans. Amer.Math.Soc., 149(1970), 569-573.

[16] W.Seaman, Helicoids of constant mean curvature and their Gauss maps, Pacific J.Math., 110(1984), 387-396.

[17] K.Voss, Über vollstandige Minimalflächen, L'Ens.Math., 10(1964), 316-317.

[18] F.Xavier, The Gauss map of a complete non-flat minimal surface cannot omit 7 points on the sphere, Ann.Math., 113('81),211-214.

Vol. 926: Geometric Techniques in Gauge Theories. Proceedings, 1981. Edited by R. Martini and E.M.de Jager. IX, 219 pages. 1982.

Vol. 927: Y. Z. Flicker, The Trace Formula and Base Change for GL (3). XII, 204 pages. 1982.

Vol. 928: Probability Measures on Groups. Proceedings 1981. Edited by H. Heyer. X, 477 pages. 1982.

Vol. 929: Ecole d'Eté de Probabilités de Saint-Flour X - 1980. Proceedings, 1980. Edited by P.L. Hennequin. X, 313 pages. 1982.

Vol. 930: P. Berthelot, L. Breen, et W. Messing, Théorie de Dieudonné Cristalline II. XI, 261 pages. 1982.

Vol. 931: D.M. Arnold, Finite Rank Torsion Free Abelian Groups and Rings. VII, 191 pages. 1982.

Vol. 932: Analytic Theory of Continued Fractions. Proceedings, 1981. Edited by W.B. Jones, W.J. Thron, and H. Waadeland. VI, 240 pages. 1982.

Vol. 933: Lie Algebras and Related Topics. Proceedings, 1981. Edited by D. Winter. VI, 236 pages. 1982.

Vol. 934: M. Sakai, Quadrature Domains. IV, 133 pages. 1982.

Vol. 935: R. Sot, Simple Morphisms in Algebraic Geometry. IV, 146 pages. 1982.

Vol. 936: S.M. Khaleelulla, Counterexamples in Topological Vector Spaces. XXI, 179 pages. 1982.

Vol. 937: E. Combet, Intégrales Exponentielles. VIII, 114 pages. 1982.

Vol. 938: Number Theory. Proceedings, 1981. Edited by K. Alladi. IX, 177 pages. 1982.

Vol. 939: Martingale Theory in Harmonic Analysis and Banach Spaces. Proceedings, 1981. Edited by J.-A. Chao and W.A. Woyczyński. VIII, 225 pages. 1982.

Vol. 940: S. Shelah, Proper Forcing. XXIX, 496 pages. 1982.

Vol. 941: A. Legrand, Homotopie des Espaces de Sections. VII, 132 pages. 1982.

Vol. 942: Theory and Applications of Singular Perturbations. Proceedings, 1981. Edited by W. Eckhaus and E.M. de Jager. V, 363 pages. 1982.

Vol. 943: V. Ancona, G. Tomassini, Modifications Analytiques. IV, 120 pages. 1982.

Vol. 944: Representations of Algebras. Workshop Proceedings, 1980. Edited by M. Auslander and E. Lluis. V, 258 pages. 1982.

Vol. 945: Measure Theory. Oberwolfach 1981, Proceedings. Edited by D. Kölzow and D. Maharam-Stone. XV, 431 pages. 1982.

Vol. 946: N. Spaltenstein, Classes Unipotentes et Sous-groupes de Borel. IX, 259 pages. 1982.

Vol. 947: Algebraic Threefolds. Proceedings, 1981. Edited by A. Conte. VII, 315 pages. 1982.

Vol. 948: Functional Analysis. Proceedings, 1981. Edited by D. Butković, H. Kraljević, and S. Kurepa. X, 239 pages. 1982.

Vol. 949: Harmonic Maps. Proceedings, 1980. Edited by R.J. Knill, M. Kalka and H.C.J. Sealey. V, 158 pages. 1982.

Vol. 950: Complex Analysis. Proceedings, 1980. Edited by J. Eells. IV, 428 pages. 1982.

Vol. 951: Advances in Non-Commutative Ring Theory. Proceedings, 1981. Edited by P.J. Fleury. V, 142 pages. 1982.

Vol. 952: Combinatorial Mathematics IX. Proceedings, 1981. Edited by E. Billington, S. Oates-Williams, and A.P. Street. XI, 443 pages. 1982.

Vol. 953: Iterative Solution of Nonlinear Systems of Equations. Proceedings, 1982. Edited by R. Ansorge, Th. Meis, and W. Törnig. VII, 202 pages. 1982.

Vol. 954: S.G. Pandit, S.G. Deo, Differential Systems Involving Impulses. VII, 102 pages. 1982.

Vol. 955: G. Gierz, Bundles of Topological Vector Spaces and Their Duality. IV, 296 pages. 1982.

Vol. 956: Group Actions and Vector Fields. Proceedings, 1981. Edited by J.B. Carrell. V, 144 pages. 1982.

Vol. 957: Differential Equations. Proceedings, 1981. Edited by D.G. de Figueiredo. VIII, 301 pages. 1982.

Vol. 958: F.R. Beyl, J. Tappe, Group Extensions, Representations, and the Schur Multiplicator. IV, 278 pages. 1982.

Vol. 959: Géométrie Algébrique Réelle et Formes Quadratiques, Proceedings, 1981. Edité par J.-L. Colliot-Thélène, M. Coste, L. Mahé, et M.-F. Roy. X, 458 pages. 1982.

Vol. 960: Multigrid Methods. Proceedings, 1981. Edited by W. Hackbusch and U. Trottenberg. VII, 652 pages. 1982.

Vol. 961: Algebraic Geometry. Proceedings, 1981. Edited by J.M. Aroca, R. Buchweitz, M. Giusti, and M. Merle. X, 500 pages. 1982.

Vol. 962: Category Theory. Proceedings, 1981. Edited by K.H. Kamps, D. Pumplün, and W. Tholen, XV, 322 pages. 1982.

Vol. 963: R. Nottrot, Optimal Processes on Manifolds. VI, 124 pages. 1982.

Vol. 964: Ordinary and Partial Differential Equations. Proceedings, 1982. Edited by W.N. Everitt and B.D. Sleeman. XVIII, 726 pages. 1982.

Vol. 965: Topics in Numerical Analysis. Proceedings, 1981. Edited by P.R. Turner. IX, 202 pages. 1982.

Vol. 966: Algebraic K-Theory. Proceedings, 1980, Part I. Edited by R.K. Dennis. VIII, 407 pages. 1982.

Vol. 967: Algebraic K-Theory. Proceedings, 1980. Part II. VIII, 409 pages. 1982.

Vol. 968: Numerical Integration of Differential Equations and Large Linear Systems. Proceedings, 1980. Edited by J. Hinze. VI, 412 pages. 1982.

Vol. 969: Combinatorial Theory. Proceedings, 1982. Edited by D. Jungnickel and K. Vedder. V, 326 pages. 1982.

Vol. 970: Twistor Geometry and Non-Linear Systems. Proceedings, 1980. Edited by H.-D. Doebner and T.D. Palev. V, 216 pages. 1982.

Vol. 971: Kleinian Groups and Related Topics. Proceedings, 1981. Edited by D.M. Gallo and R.M. Porter. V, 117 pages. 1983.

Vol. 972: Nonlinear Filtering and Stochastic Control. Proceedings, 1981. Edited by S.K. Mitter and A. Moro. VIII, 297 pages. 1983.

Vol. 973: Matrix Pencils. Proceedings, 1982. Edited by B. Kågström and A. Ruhe. XI, 293 pages. 1983.

Vol. 974: A. Draux, Polynômes Orthogonaux Formels - Applications. VI, 625 pages. 1983.

Vol. 975: Radical Banach Algebras and Automatic Continuity. Proceedings, 1981. Edited by J.M. Bachar, W.G. Bade, P.C. Curtis Jr., H.G. Dales and M.P. Thomas. VIII, 470 pages. 1983.

Vol. 976: X. Fernique, P.W. Millar, D.W. Stroock, M. Weber, Ecole d'Eté de Probabilités de Saint-Flour XI - 1981. Edited by P.L. Hennequin. XI, 465 pages. 1983.

Vol. 977: T. Parthasarathy, On Global Univalence Theorems. VIII, 106 pages. 1983.

Vol. 978: J. Ławrynowicz, J. Krzyż, Quasiconformal Mappings in the Plane. VI, 177 pages. 1983.

Vol. 979: Mathematical Theories of Optimization. Proceedings, 1981. Edited by J.P. Cecconi and T. Zolezzi. V, 268 pages. 1983.

Vol. 980: L. Breen. Fonctions thêta et théorème du cube. XIII, 115 pages. 1983.

Vol. 981: Value Distribution Theory. Proceedings, 1981. Edited by I. Laine and S. Rickman. VIII, 245 pages. 1983.

Vol. 982: Stability Problems for Stochastic Models. Proceedings, 1982. Edited by V. V. Kalashnikov and V. M. Zolotarev. XVII, 295 pages. 1983.

Vol. 983: Nonstandard Analysis-Recent Developments. Edited by A.E. Hurd. V, 213 pages. 1983.

Vol. 984: A. Bove, J. E. Lewis, C. Parenti, Propagation of Singularities for Fuchsian Operators. IV, 161 pages. 1983.

Vol. 985: Asymptotic Analysis II. Edited by F. Verhulst. VI, 497 pages. 1983.

Vol. 986: Séminaire de Probabilités XVII 1981/82. Proceedings. Edited by J. Azéma and M. Yor. V, 512 pages. 1983.

Vol. 987: C. J. Bushnell, A. Fröhlich, Gauss Sums and p-adic Division Algebras. XI, 187 pages. 1983.

Vol. 988: J. Schwermer, Kohomologie arithmetisch definierter Gruppen und Eisensteinreihen. III, 170 pages. 1983.

Vol. 989: A. B. Mingarelli, Volterra-Stieltjes Integral Equations and Generalized Ordinary Differential Expressions. XIV, 318 pages. 1983.

Vol. 990: Probability in Banach Spaces IV. Proceedings, 1982. Edited by A. Beck and K. Jacobs. V, 234 pages. 1983.

Vol. 991: Banach Space Theory and its Applications. Proceedings, 1981. Edited by A. Pietsch, N. Popa and I. Singer. X, 302 pages. 1983.

Vol. 992: Harmonic Analysis, Proceedings, 1982. Edited by G. Mauceri, F. Ricci and G. Weiss. X, 449 pages. 1983.

Vol. 993: R. D. Bourgin, Geometric Aspects of Convex Sets with the Radon-Nikodým Property. XII, 474 pages. 1983.

Vol. 994: J.-L. Journé, Calderón-Zygmund Operators, Pseudo-Differential Operators and the Cauchy Integral of Calderón. VI, 129 pages. 1983.

Vol. 995: Banach Spaces, Harmonic Analysis, and Probability Theory. Proceedings, 1980–1981. Edited by R.C. Blei and S.J. Sidney. V, 173 pages. 1983.

Vol. 996: Invariant Theory. Proceedings, 1982. Edited by F. Gherardelli. V, 159 pages. 1983.

Vol. 997: Algebraic Geometry – Open Problems. Edited by C. Ciliberto, F. Ghione and F. Orecchia. VIII, 411 pages. 1983.

Vol. 998: Recent Developments in the Algebraic, Analytical, and Topological Theory of Semigroups. Proceedings, 1981. Edited by K.H. Hofmann, H. Jürgensen and H. J. Weinert. VI, 486 pages. 1983.

Vol. 999: C. Preston, Iterates of Maps on an Interval. VII, 205 pages. 1983.

Vol. 1000: H. Hopf, Differential Geometry in the Large, VII, 184 pages. 1983.

Vol. 1001: D.A. Hejhal, The Selberg Trace Formula for PSL(2, ℝ). Volume 2. VIII, 806 pages. 1983.

Vol. 1002: A. Edrei, E.B. Saff, R.S. Varga, Zeros of Sections of Power Series. VIII, 115 pages. 1983.

Vol. 1003: J. Schmets, Spaces of Vector-Valued Continuous Functions. VI, 117 pages. 1983.

Vol. 1004: Universal Algebra and Lattice Theory. Proceedings, 1982. Edited by R.S. Freese and O.C. Garcia. VI, 308 pages. 1983.

Vol. 1005: Numerical Methods. Proceedings, 1982. Edited by V. Pereyra and A. Reinoza. V, 296 pages. 1983.

Vol. 1006: Abelian Group Theory. Proceedings, 1982/83. Edited by R. Göbel, L. Lady and A. Mader. XVI, 771 pages. 1983.

Vol. 1007: Geometric Dynamics. Proceedings, 1981. Edited by J. Palis Jr. IX, 827 pages. 1983.

Vol. 1008: Algebraic Geometry. Proceedings, 1981. Edited by J. Dolgachev. V, 138 pages. 1983.

Vol. 1009: T. A. Chapman, Controlled Simple Homotopy Th Applications. III, 94 pages. 1983.

Vol. 1010: J.-E. Dies, Chaînes de Markov sur les permuta 226 pages. 1983.

Vol. 1011: J.M. Sigal. Scattering Theory for Many-Body Mechanical Systems. IV, 132 pages. 1983.

Vol. 1012: S. Kantorovitz, Spectral Theory of Banach Spa ators. V, 179 pages. 1983.

Vol. 1013: Complex Analysis – Fifth Romanian-Finnish Part 1. Proceedings, 1981. Edited by C. Andreian Cazacu, N M. Jurchescu and I. Suciu. XX, 393 pages. 1983.

Vol. 1014: Complex Analysis – Fifth Romanian-Finnish Part 2. Proceedings, 1981. Edited by C. Andreian Cazacu, N. M. Jurchescu and I. Suciu. XX, 334 pages. 1983.

Vol. 1015: Equations différentielles et systèmes de Pfaff champ complexe – II. Seminar. Edited by R. Gérard et J. F V, 411 pages. 1983.

Vol. 1016: Algebraic Geometry. Proceedings, 1982. Edite Raynaud and T. Shioda. VIII, 528 pages. 1983.

Vol. 1017: Equadiff 82. Proceedings, 1982. Edited by H. W. K and K. Schmitt. XXIII, 666 pages. 1983.

Vol. 1018: Graph Theory, Łagów 1981. Proceedings, 1981. E M. Borowiecki, J. W. Kennedy and M. M. Sysło. X, 289 page

Vol. 1019: Cabal Seminar 79–81. Proceedings, 1979–81. E A. S. Kechris, D. A. Martin and Y. N. Moschovakis. V, 284 page

Vol. 1020: Non Commutative Harmonic Analysis and Lie Proceedings, 1982. Edited by J. Carmona and M. Vergne. pages. 1983.

Vol. 1021: Probability Theory and Mathematical Statistic ceedings, 1982. Edited by K. Itô and J.V. Prokhorov. VIII, 747 1983.

Vol. 1022: G. Gentili, S. Salamon and J.-P. Vigué. Geometry S "Luigi Bianchi", 1982. Edited by E. Vesentini. VI, 177 pages.

Vol. 1023: S. McAdam, Asymptotic Prime Divisors. IX, 118 1983.

Vol. 1024: Lie Group Representations I. Proceedings, 1982 Edited by R. Herb, R. Lipsman and J. Rosenberg. IX, 369 pages

Vol. 1025: D. Tanré, Homotopie Rationnelle: Modèles de Quillen, Sullivan. X, 211 pages. 1983.

Vol. 1026: W. Plesken, Group Rings of Finite Groups Over Integers. V, 151 pages. 1983.

Vol. 1027: M. Hasumi, Hardy Classes on Infinitely Connecte mann Surfaces. XII, 280 pages. 1983.

Vol. 1028: Séminaire d'Analyse P. Lelong – P. Dolbeault – H. Années 1981/1983. Edité par P. Lelong, P. Dolbeault et H. VIII, 328 pages. 1983.

Vol. 1029: Séminaire d'Algèbre Paul Dubreil et Marie-Paule Ma Proceedings, 1982. Edité par M.-P. Malliavin. V, 339 pages. 1

Vol. 1030: U. Christian, Selberg's Zeta-, L-, and Eisensteins XII, 196 pages. 1983.

Vol. 1031: Dynamics and Processes. Proceedings, 1981. Edit Ph. Blanchard and L. Streit. IX, 213 pages. 1983.

Vol. 1032: Ordinary Differential Equations and Operators ceedings, 1982. Edited by W. N. Everitt and R. T. Lewis. XV, 521 p 1983.

Vol. 1033: Measure Theory and its Applications. Proceedings, Edited by J. M. Belley, J. Dubois and P. Morales. XV, 317 pages.

Vol. 1034: J. Musielak, Orlicz Spaces and Modular Spac 222 pages. 1983.